Fatigue strength of welded structures

Fatigue strength of welded structures

SECOND EDITION

S J MADDOX
BSc (Eng), PhD, DIC, CEng, FWeldI

ABINGTON PUBLISHING

Woodhead Publishing Ltd in association with The Welding Institute
Cambridge England

Published by Abington Publishing,
Woodhead Publishing Ltd, Abington Hall, Abington,
Cambridge CB1 6AH, England

First published 1969, The Welding Institute
Second edition 1991, Abington Publishing
Reprinted 1994

British Library Cataloguing in Publication Data

Maddox S J (Stephen John) 1944–
Fatigue Strength of Welded Structures, 2nd ed
1. Joints (structures) Analysis
I. Title II. Richards K G (Kenneth George) 1932–
624.171

ISBN 1 85573 013 8

Designed by Geoff Green (text) and Chris Feely (jacket),
typeset by Bookens Ltd, Baldock and printed by
St Edmundsbury Press Ltd, Bury St Edmunds, Suffolk.

Contents

Part 2 Design rules

Foreword

Knowledge of the fatigue behaviour of welded structures has continued to be gathered at some pace, from observations of service performance as well as from extensive and important research studies, during the twenty years or so since the first account of the problem was published in booklet form by The Welding Institute. As a result, a good deal more can now be said, especially by way of guidance on the measures to be taken in design to reduce the likelihood of premature or rapidly progressive fatigue cracking.

The need for such guidance remains a paramount requirement. Welded structures are capable of fulfilling a wide range of demanding design objectives, including successful service under loading conditions that can produce fatigue damage, given the opportunity. Effective resistance to these conditions depends to a very large extent, however, upon a design approach which embraces a good understanding of the nature of the fatigue problem and a sure grasp of the way in which the effects of welding must be taken into account. It is a very poor and costly alternative to seek immunity by the arbitrary increase of scantlings and weight.

In consideration of the need to present a full appreciation of the problem before setting out design procedures and related material, this book is divided into two sections, identified as Part 1 and Part 2, and represents a considerably revised and extended version of the original edition published under the same title.

The first part, although an essential foundation to the employment of design rules, offers more or less self-contained reading in taking the story from an analysis of cause and effect up to a statement of what are virtually standard fatigue strength data for arc welded joints in steel and aluminium alloy. In this respect, Part 1 provides an account which is readily incorporated into the teaching of student engineers and will be seen in this context to be of particular value in putting into important perspective one or two of the supposedly classical rules of fatigue of metals, hopefully to the benefit of future engineering design.

Part 2 presents design rules for welded structures destined for exposure to fatigue loading. The central feature of the revised rules is an amended classification of welded joints in steel coupled with a completely new statement of design stresses for each class. These revisions have stemmed from a statistical analysis of a large collection of experimental data, leading to the production of three design S-N curves per class (for constant amplitude loading), i.e. the mean S-N curve and curves for the mean minus one and mean minus two standard deviations of log N. Design stresses can therefore be related to selected survival limits, and the general approach is one shown to be applicable also to aluminium alloys.

Practical aspects of the application of the rules for design in steel are discussed, including consideration of sources of additional stress concentration and their treatment, the effect of plate thickness, the influence of thermal stress relief, and ways in which high strength steels might be used to advantage. The importance of attention to detail in fatigue design is illustrated by reference to actual service failures.

Although not advocated as substitutes for good design, post-weld fatigue strength improvement methods have their uses. The principles upon which they rely, their practical application, and resulting improvements in fatigue life are described. Also given is an account of the way in which, in one of the most important recent advances, the problem of the significance of weld imperfections has been reduced to an engineering critical assessment procedure for the specification of acceptance limits on a fitness-for-purpose basis. The procedure is quite capable of prompting deep consideration of flaw detection and measurement, but it will be understood that this was always going to be so with any system of acceptance, other than the somewhat vain specification 'defect free'.

Finally, there is a major addition to the book in the form of an introduction to the fracture mechanics analysis of fatigue behaviour. Fracture mechanics provides a powerful means of assessing cracks and cracklike flaws, and in particular can closely estimate fatigue crack propagation rates, given that stress input conditions are reasonably well known. Many design cases will be recognised where this ability can be employed to make possible a crucial check on the cyclic life required for crack growth to a size that might, for example, cause leakage or represent a critical condition in respect of fracture resistance.

In summary, Part 2 gives a survey of design procedures and guidelines which is as up to date as possible. It will not, of course, provide the designer of welded structures with answers to every conceivable question. Research is still necessary, for the reason that design aims, materials, and construction methods continuously change. It is believed, however, that the body of knowledge presented will provide substantial aid towards successful contemporary design.

R P Newman

Part 1

1 Welded joints under fatigue loading

Chapter 1

Fatigue failure

In the context of engineering, fatigue is the process by which a crack can form and then grow under repeated or fluctuating loading. Extensive fatigue cracking may exist in some types of structure without seriously affecting their load-carrying capacity, although 'fatigue failure' is said to have occurred. On the other hand, positive damage such as sudden collapse, excessive deflection, or leakage of a container can be induced if the crack is permitted to propagate to some critical size.

The magnitude of the loading required to produce fatigue cracking in a component may be much less than that needed to break the component in a single application of load. In particular, it may be much less than the load corresponding to the allowable static design stress. Under normal conditions allowable static design stresses are related to the yield strength of the material: for example, steel structures may be designed to stresses of approximately two-thirds of the yield strength. This gives an adequate margin against the onset of yield and a bigger margin against ultimate failure. In other words, there is a factor of safety. This is true provided it is established that the structure will be subjected only to static loading, i.e. the load (and stress) remain constant with time. Safety may also depend on other factors such as liability to brittle fracture, instability, corrosion, and so on, but these are excluded from the present discussion.

In practice few structures are subjected to purely static loading and, in general, some fluctuations in stress arise. These changes in stress may range from a simple cyclic fluctuation to a completely random variation. Although they may be within the limit of the allowable static stress, the structure can no longer be regarded as safe without first considering the possibility of fatigue cracking.

It is important to recognise loading which can give rise to fatigue, that is fatigue loading. Any variation in the loading experienced by a member, no matter how small or infrequent, should be considered as potential fatigue loading. Fatigue failures have occurred in as little as a few hundred, and as many as several millions of, load changes. Variations in live load applied during normal operations, pressure changes, vibrations, temperature fluctuations and wave and wind forces are some of the ways in which fluctuations of the working stress arise in a structure. Welded structures affected include, among many others, bridges, cranes, excavators, ships, vehicles, masts, offshore structures, process and manufacturing plant, and pipelines. In some instances, notably steel bridges, steel cranes, offshore steel structures, and pressure vessels, there are national standards containing design stresses relating to the avoidance of fatigue failure.

Since fatigue cracking can and generally does arise under elastic nominal stresses, it is not surprising that fatigue damage is highly localised and fatigue cracks are difficult to see. When failure of a ductile material is brought about by a gradually increasing load from zero to a maximum, rupture of the material is preceded by considerable deformation which can be seen clearly without the aid of measuring instruments. In a tensile test on mild steel, for example, an elongation of at least 20% would be expected. By comparison, fatigue-cracked components exhibit no visible signs of deformation of the material, Fig. 1, unless the extent of fatigue cracking is such that yielding occurs on the uncracked section.

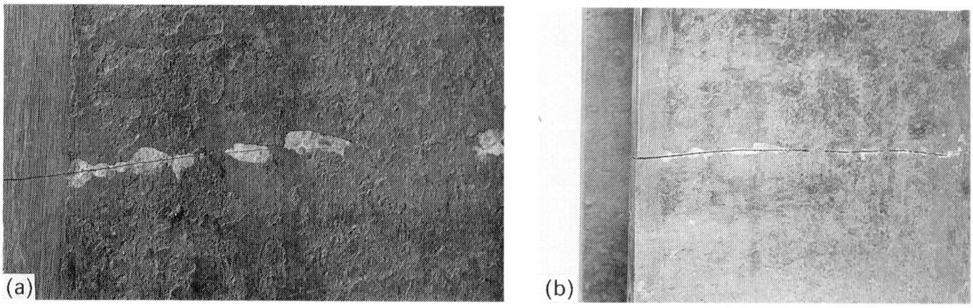

1 Appearance of fatigue crack at two stages of propagation through mild steel plate showing absence of yielding.

2 Typical fatigue fracture in steel showing striated surface.

A fatigue crack starts as a very small fissure which grows in size as the repeated stressing is maintained until it extends through a considerable part of the material. It may be very difficult to detect its presence during this period, as illustrated later.

The fracture surface of a fatigue crack has a characteristic appearance. It usually possesses a smooth, matt surface and frequently shows concentric rings or 'beach marks', spreading out from discrete points where the fracture was initiated, Fig. 2. These markings are often a feature of service failures associated with non-uniform loading conditions, intervals of rest and differential corrosion. The fracture surface also tends to become less smooth as the rate of propagation of the crack increases.

On a microscopic scale the fracture also often shows 'striations' which usually represent the crack front position after successive stress cycles, Fig. 3.

The rate of propagation of the crack can vary considerably depending on the material, the type of component and the nature of the applied stressing. In a highly redundant structure a crack may initiate in an area of high stress and propagate to a lowly stressed area where, if there are alternative load paths, it may stop altogether. In other members the formation of a crack will increase the net-section stress because of the reduction in cross-section, so that the rate of propagation

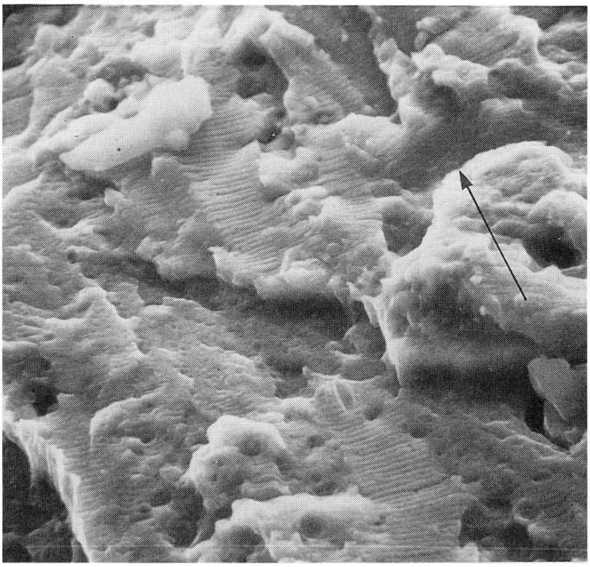

3 Microscopic striations on fatigue fracture surface in aluminium alloy. Direction of crack growth indicated (×1350).

will increase as the crack progresses until the load on the remaining ligament produces final rupture.

Fatigue loading

As already mentioned, fluctuating stresses can arise in a number of ways and the resulting stress-time variations may be quite different. Consider two simple examples:

Simply supported crane girder with constant travelling load, W, producing, for each load traverse, a maximum bending moment of WL/4 at section X-X, Fig. 4.

Service conditions:
Load W, constant
Two crane passes/hr, 12hr/day, 240 days/yr

In a service life of forty years the girder accumulates:
2 × 12 × 240 × 40
= 230 400 repetitions (cycles) of bending moment WL/4

Rotating shaft with overhung flywheel producing bending moment, WL, at point X, Fig. 5

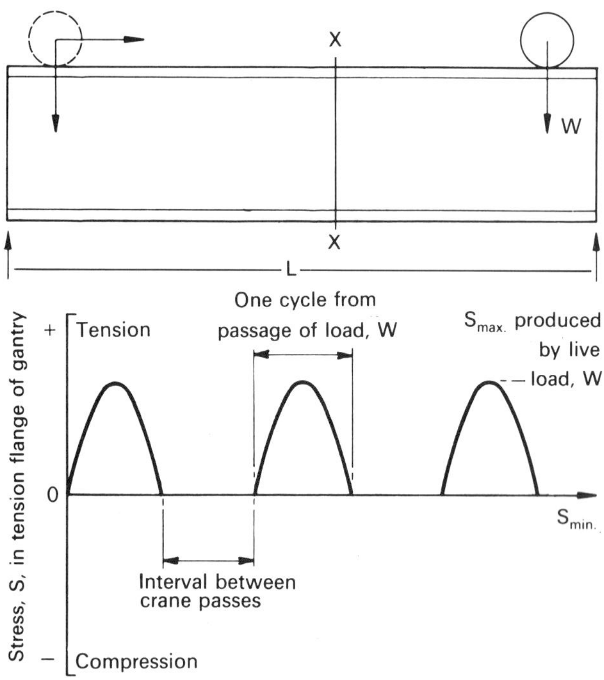

4 Typical fatigue loading of crane girder.

Service conditions:
Load, W, constant
Shaft rotates at 250 rev/min, 8hr/day, 300 days/yr

In a service life of forty years the shaft accumulates
$250 \times 60 \times 8 \times 300 \times 40$
= 1440M repetitions (cycles) of bending moment, WL

The two cases make a striking comparison in terms of the numbers of load, bending moment and stress cycles accumulated over the same period of time. This results from the basic difference of working schedule between the crane gantry girder and the shaft.

In the first example, as the load passes across the crane girder the stress in the lower flange at X–X changes from $S_{min} = 0$ (ignoring self-weight stresses) to a tension stress of $S_{max} = \frac{WL}{4Z}$, where Z is section

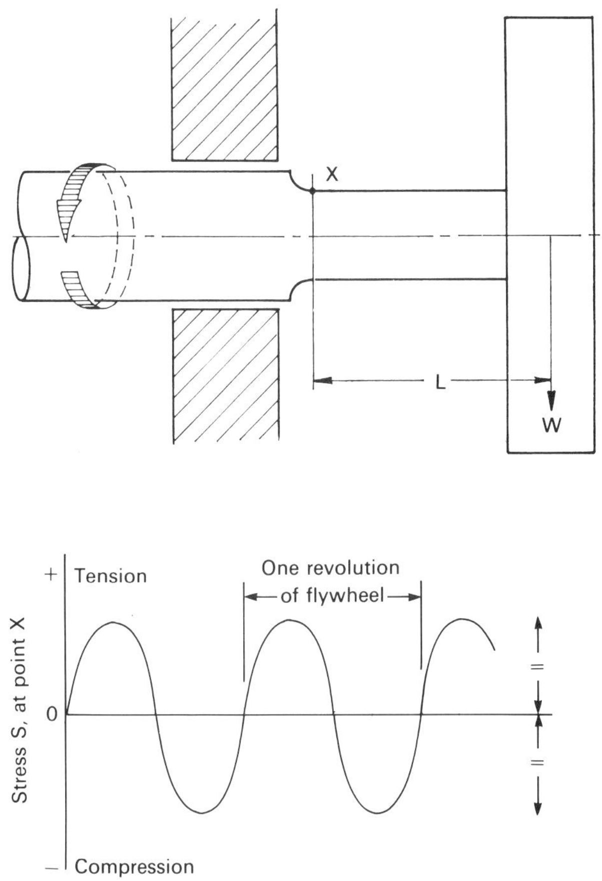

5 Fatigue loading in rotating shaft.

modulus. The bending stresses in the second example alternate between equal tension and compression for each revolution of the flywheel, that is $S_{min} = - S_{max}$. Thus, although both examples show simple cyclic stressing, the mean stress level in the first is a tension stress, $\frac{1}{2}S_{max}$, and in the second the mean stress is zero. Other examples could show that mean stress might be negative or compressive.

Conventionally, a fluctuating stress is represented as a series of peaks and troughs connected by a sine wave, Fig. 6. The diagram shows what is termed 'constant amplitude' loading, each cycle having identical values of S_{max} and S_{min}. This simple form of cyclic stress is often used in testing and the fatigue strength data for welded joints quoted later refer to constant amplitude conditions which, in turn, can be described by several parameters, i.e:

Upper limit stress, S_{max}
Lower limit stress, S_{min}
Mean stress, $S_m = \frac{1}{2} (S_{max} + S_{min})$
Stress range, $S_r = S_{max} - S_{min}$

The use of this terminology varies, depending upon the kind of problem under consideration. Thus, if S_{min} is always zero in a particular case, data may be given in terms of S_{max}. Conversely, if S_m is always zero, as in the example of the shaft, data will be given in the form $\pm S_a$, where S_a, the stress amplitude, is half of the stress range. $S_m \pm S_a$ is often used when S_m is other than zero.

It is necessary for general design purposes to have fatigue data for positive and negative values of S_{max}, S_{min}, and S_m to cater for stress conditions of the kind shown in Fig. 7. In this respect, use is made of a characterising parameter, i.e. stress ratio, R; quite simply:

Stress ratio, R = minimum stress/maximum stress

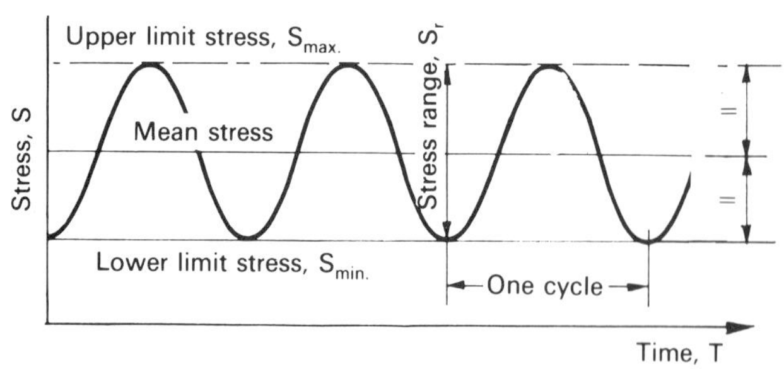

6 Terminology used to describe constant amplitude fluctuating stress.

There are two ways of defining minimum and maximum stress and therefore the calculation of R. In both, tensile stresses are taken as positive and compressive stresses as negative; the two methods are:

$$R = \frac{\text{arithmetical minimum stress}}{\text{arithmetical maximum stress}} \text{ or } R = \frac{\text{algebraic minimum stress}}{\text{algebraic maximum stress}}$$

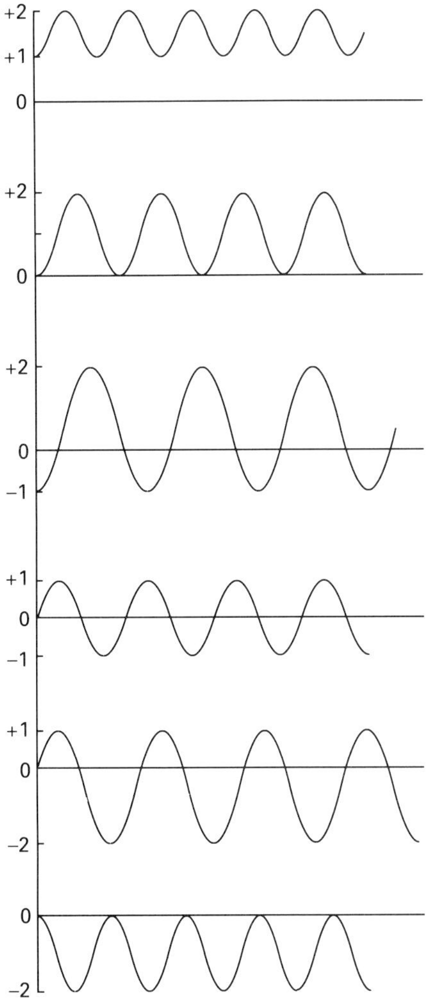

Stress ratio	
Arithmetical minimum stress	Algebraic minimum stress
$\frac{+1}{+2} = +0{\cdot}5$	$\frac{+1}{+2} = +0{\cdot}5$
$\frac{0}{+2} = 0$	$\frac{0}{+2} = 0$
$\frac{-1}{+2} = -0{\cdot}5$	$\frac{-1}{+2} = -0{\cdot}5$
$\frac{-1}{+1} = -1$	$\frac{-1}{+1} = -1$
$\frac{-2}{+1} = -2$	$\frac{+1}{-2} = -0{\cdot}5$
$\frac{-2}{0} = \infty$	$\frac{0}{-2} = 0$

7 Definition of stress ratio.

In the first calculation S_{min} is the stress at the trough of the waveform; in the second it is the smaller stress, irrespective of whether it is positive or negative. Figure 7 clarifies these considerations and also shows that the algebraic method can give the same value of R for different waveforms. This ambiguity is accepted in design practice. Take, for example, the two loading conditions which each gives R=−0.5: S_{max} (algebraic) is +2 and −2 respectively; thus, S_{max} automatically defines the major load to be designed for, whether tensile or compressive. The convention is to refer to R=−0.5 (S_{max} tensile), or R=−0.5 (S_{max} compressive). However, to simplify further discussion, the arithmetical determination of R is used in this book.

Each stress cycle contributes to the initiation and growth of fatigue cracks. The number of cycles needed to produce fatigue failure, termed the fatigue life, depends on the type of loading. In general it is tensile stresses which produce fatigue damage, so that a stress variation of a given range which is partly compressive will be less damaging than one which is wholly tensile, and consequently the fatigue life will be greater. Furthermore, the damaging effect of a fully tensile cyclic stress range tends to increase as the mean stress or stress ratio increases. For a given stress range, a cyclic stress with R=+0.5 will therefore be more damaging than one with R=0 and will result in a shorter fatigue life. It will be shown, however, that these basic principles do not always adequately predict the fatigue behaviour of welded joints, especially those in the 'as-welded' condition, containing residual stresses.

Loading conditions in service can introduce a number of modifications of the simple waveform illustrated in Fig. 6. The variations illustrated in Fig. 8 relate to waveform shape, frequency, and periods of rest. Each of these three variables is important in given circumstances, for example when it is necessary to consider additional environmental factors such as elevated temperature or corrosive conditions. High temperature fatigue and corrosion fatigue are, however, special aspects of the general problem and if, as is often the situation, they can be neglected, the variables noted assume much less significance. Many design codes covering 'orthodox' fields of engineering discount the effect of loading frequency and periods of rest, and the waveforms shown in Fig. 8 would be taken as equally damaging because the stress range is the same in each case.

Fatigue test data

Referring to the examples of cyclically loaded components discussed above it will be noted that, although each was required to give a service life of forty years, the number of cycles of load which must be sustained in that life were very different. The design of the crane girder may be based on a limited number of cycles, but the shaft must be designed to give what is virtually an infinite fatigue life in terms of the number of loading cycles. Not surprisingly, the stresses which could be applied in service to give these lives are quite different. The expected fatigue endurance of a component subjected to a particular cyclic stress is obtained from an S–N curve for that geometry of component.

By testing a series of identical specimens in the laboratory it is possible to establish the relationship between nominal stress range and number of cycles to failure for any particular type of loading. The results from such a series of tests may be used to produce the traditional S–N curve. A series of such curves may need to be produced for the different types of loading which can occur, that is with different levels of mean stress. By using logarithmic scales, the test results fall close to a straight line, Fig. 9, which extends down to an 'endurance limit', S_0, below which

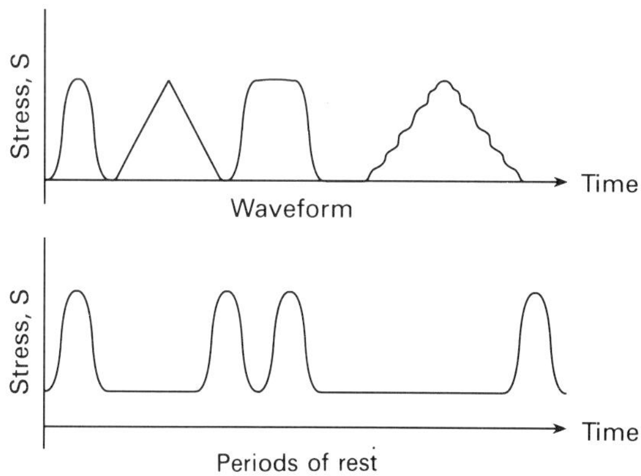

8 Variations of waveform.

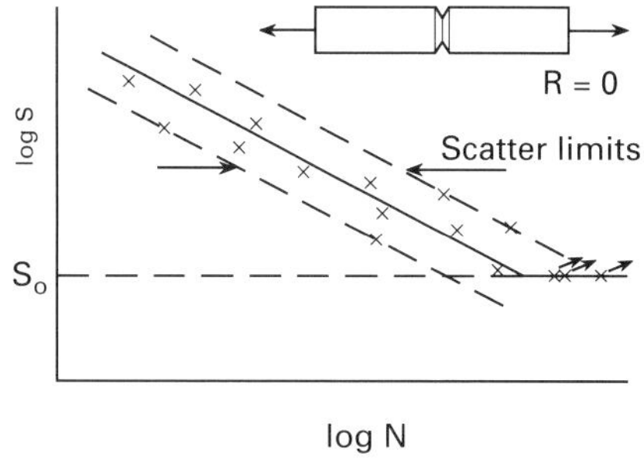

9 Fatigue test data and S–N curve.

fatigue damage is negligible and the fatigue life is therefore effectively infinite. In the general context of the problem, fatigue strength is often quoted in terms of the endurance limit, and indeed that stress would be relevant to the shaft design. However, it is rarely economic to design welded structures to give infinite life because the allowable stress would be so low; instead, the stress corresponding to the estimated number of load cycles which will occur in the required life is used in design. It must be recognised that this means the welded component will contain a slowly growing crack, perhaps for most of its life.

It is important to use identical test specimens to obtain S–N data because, as will be seen later, geometry has a major influence on fatigue strength. In practice, variations in geometry occur; these, and to a lesser extent material variations, contribute to scatter in fatigue test data, Fig. 9. Scatter of results leads to uncertainties about the magnitude of cyclic stress which can be applied to a component for which a particular fatigue life is required, and statistics are normally used to interpret test data. The mean S–N curve represents a 50/50 chance of failure and a component designed on the basis of the mean curve has an equal chance of failure or survival. In design, a greater chance of survival is usually required and the design S–N curve falls below the mean. This and other aspects of design practice are fully discussed in Pt 2 of this book which, in particular, considers the problem of design when random (varying stress range) rather than constant amplitude loading has to be taken into account.

When comparing the fatigue strengths of different components subjected to similar loading, it is convenient to express fatigue strength in terms of the stresses corresponding to particular lives, for example 10^5, 10^6, and 10^7 cycles on the mean S–N curve. The choice of reference life is quite arbitrary. Traditionally, 2×10^6 cycles has been used, and indeed some design codes refer to their S–N curves in terms of the corresponding stress range. For these reasons, in the discussion later in this book the fatigue strengths of welded joints in steel at 2×10^6 cycles are taken as a basis for comparison. It should be noted that a comparison of different components based on the fatigue strength corresponding to a given life will be misleading if the slopes of the S–N curves of the two components are significantly different. The ratio of fatigue strength at one life would then be different from that at another. However, the S–N curves for most welded joints are in fact parallel and comparisons of fatigue strengths at a nominated life can be made.

Fatigue testing machines, particularly those needed to test relatively large specimens, apply cyclic loads at frequencies typically in the range 5–15Hz, Fig. 10. Clearly this means that the generation of fatigue test data can be a lengthy process, even for one type of specimen and one particular stress ratio. Indeed, it takes 8–24 days to apply 10^7

cycles. For this reason most test series are normally confined to fatigue lives of less than 10^7 cycles. In addition, the number of test series carried out at different stress ratios is limited and results corresponding to ratios for which test data are not available are obtained by interpolation.

The process of interpolation involves the use of a diagram which shows the relationship between maximum (S_{max}) and minimum (S_{min}) stress in stress cycles which each produces failure at a specified life. In one such diagram, the Smith diagram, Fig. 11, S_{max} and S_{min} are plotted against mean stress, S_m. Empirical rules exist which, for example, would predict values of S_{max} at points B_1 and C_1, given that S_{max} at

10 Fatigue testing butt welded specimen.

A_1 was known. Experimental work can, however, determine principal values such as A_1–A_2 (R=−1), B_1–B_2 (R=0), and C_1–C_2 (R=+0.5). The vertical distance between each pair of points represents stress range, S_r, and the diagram indicates how, for the specified life, S_r varies with S_m.

It is evident that the Smith diagram has the disadvantage that information relating to other than one particular life cannot be included with complete clarity. Fatigue strengths for different lives can nevertheless

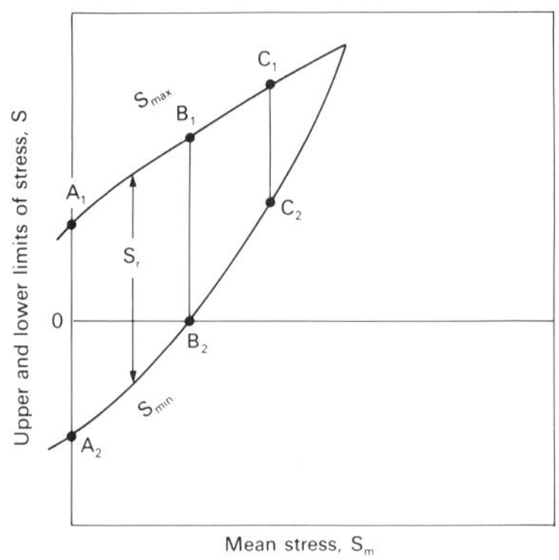

11 Smith diagram showing stress conditions corresponding to particular fatigue life.

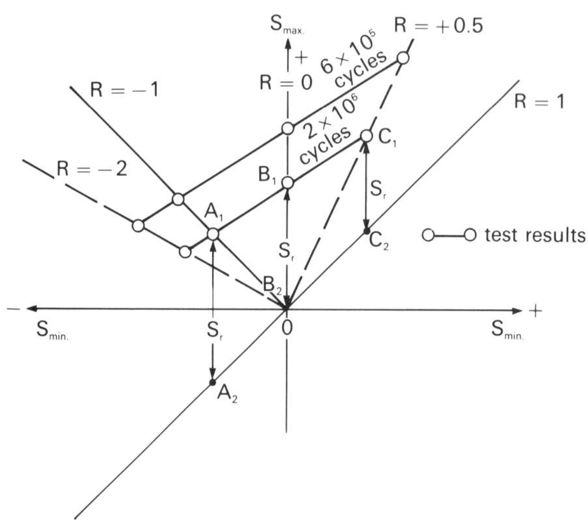

12 Stress conditions corresponding to particular fatigue lives.

be compared in a different form of diagram. Fig. 12, in which S_{max} is plotted against S_{min}. The fatigue strength values to be entered will be for a given component, for selected values of R, and for fatigue lives of interest (in the example shown, for 6×10^5 and 2×10^6 cycles); each set of experimental points is then joined to give a reference curve for the life concerned. Values of S_r are determined as indicated in the diagram, these again referring to R=−1 (A_1-A_2), =0 (B_1-B_2), =+0.5 (C_1-C_2), in each case for 2×10^6 cycles; note that the radiating R lines are so arranged that each represents a fixed ratio of S_{min} to S_{max}.

As mentioned earlier, the S–N curve gives the relationship between the applied cyclic stress range and fatigue life. For design purposes the latter must be determined first, by careful consideration of the nature and intended length of service of the components or structure concerned. Only then can the permissible stress range be established. Engineering design is, however, invariably affected by constraints so that, for example, an economic life relating to capital cost must be chosen and, if necessary, fatigue considerations must be made to comply. There is, nevertheless, some scope in design, especially in the design of welded structures, to influence the S–N relationship (really to influence the choice of S_r for a fixed value of N) by controlling the configuration of the component or structure and the detailed geometry of the joints. This aspect represents the third important factor which influences fatigue strength. Performance under the same applied stress conditions can vary widely from one component to the next, depending upon configuration and joint geometry.

Stress concentrations

A basic question not yet answered is why fatigue cracks initiate at all when the applied stress is less than the ultimate strength of the component. The answer lies in the incidence of stress concentrations. Any discontinuity in a stressed member introduces a stress concentration, that is a region where, locally, the level of stress is raised above the average. This effect can be visualised in terms of the stress flow patterns in stressed plates containing discontinuities, a hole and a notch, and the actual stress distributions across the net sections, Fig. 13.

These diagrams show that the stress is not uniformly distributed, as it is in plates without a hole or notch. Not only is the stress distribution not uniform but, adjacent to the edge of the hole or root of the notch, it is considerably higher in magnitude than the average stress based on the net cross-sectional area. The existence of these stress concentrations can be demonstrated both by calculation and by experimental techniques such as photoelasticity, Fig. 14. For example, it is possible for the peak stress at the edge of a hole in a tensile stress field to be three times the average stress on the net cross-sectional area of the plate (depending on the relative dimensions of W and d), giving a stress concentration factor (peak stress/average stress) K_t=3.0.

The effect of stress concentrations of these types on the fatigue strength of structural steel plates is shown in S–N curves, Fig. 15, obtained for plain plate, the same plate with a small hole drilled through the centre, and finally the plate with small V notches cut in both edges, all subjected to repeated tension loading, R=0. There is a considerable difference in the fatigue strengths of the three types of specimen, the lowest strength being associated with the plate containing the highest stress concentration, the V notch, which is the most abrupt discontinuity in the stress path.

The above stress concentrations result from features which reduce

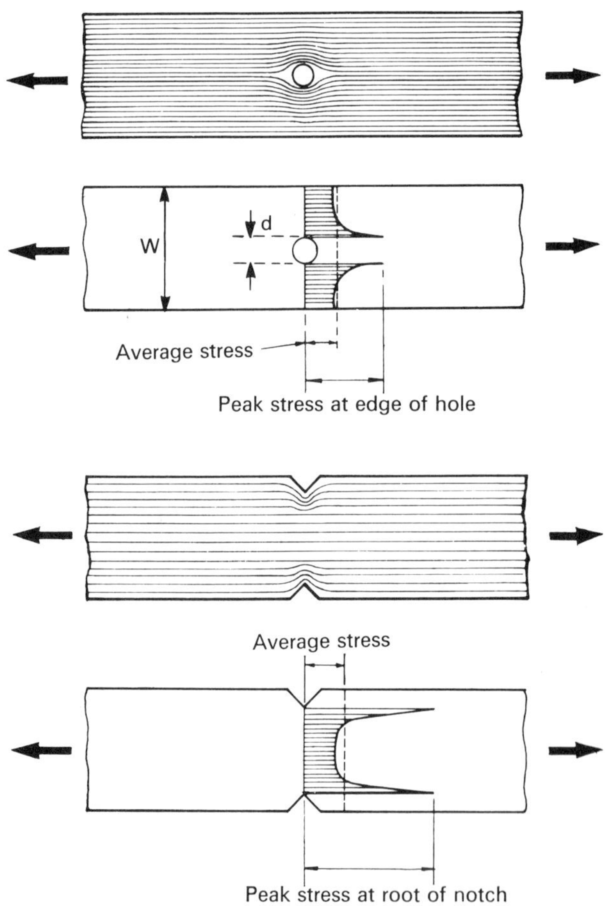

13 Stress concentrations.

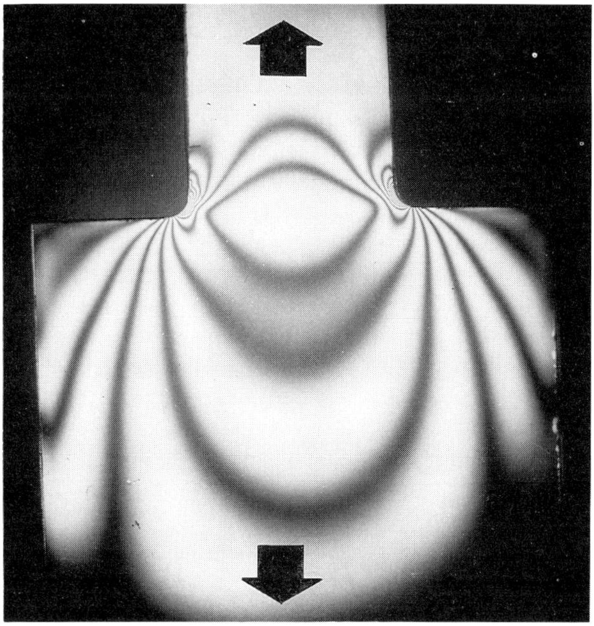

14 Photoelastic fringe pattern indicating stress concentration at change in section under axial loading.

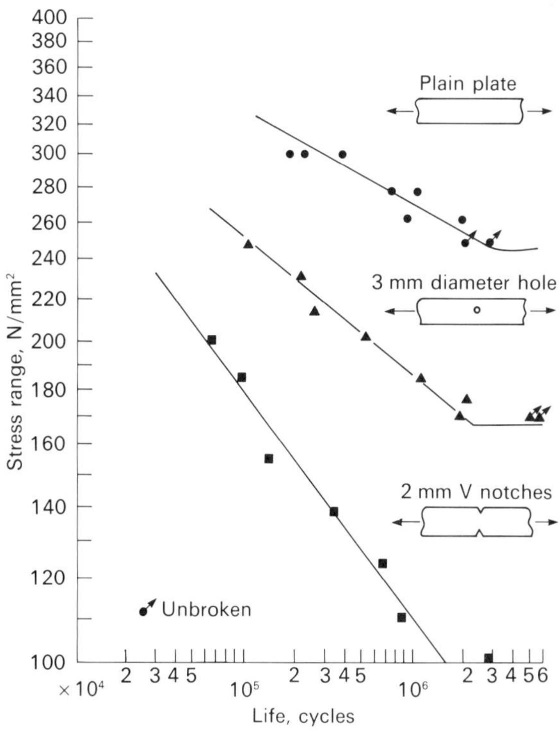

15 Effect of stress concentrations on fatigue strength of mild steel plate.

the cross-section of the plate. However, the condition of stress concentration arises as a consequence of discontinuity or change of shape and not the reduction in cross-section, so that even local increases of section can introduce stress concentrations. Note in this respect that section changes must introduce disturbances in the flow of stress before they cause stress concentration. This means that their orientations with respect to the direction of stress are important. The increase in section shown in Fig. 16 does not introduce a stress concentration with respect to P_x, only to P_y.

Stress concentrations similar to those illustrated exist at drilled holes, at the roots of screw threads, under bolt heads, and at shoulders on shafts, and are the reason why many service fatigue failures occur at details of this type. Under static loading a stress concentration in a ductile material has no effect on the strength of the component concerned because the volume of highly stressed material in the field of stress concentration is so small compared with the amount of material at nominal stress. However, under cyclic loading that small volume of material is repeatedly highly stressed and eventually damage occurs in the form of cracking. Once a crack is present the stress concentration intensifies and continued propagation of the crack by the process of repeated high stressing of the small volume of material at the crack tip becomes easier and easier and the rate of crack propagation gradually increases.

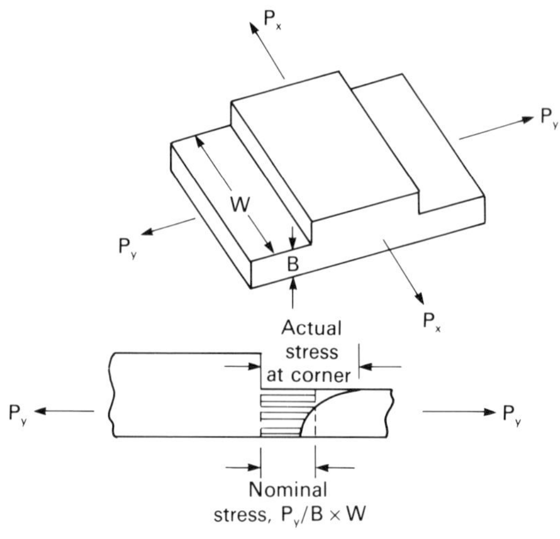

16 Stress concentration at increase of section.

Chapter 2

Significant features of welds in relation to fatigue

If a component or structure will be subjected to fluctuating loading in service, the avoidance of fatigue failure is likely to be a factor which limits the design stresses. This is particularly true of welded components because their fatigue strengths can be much lower than those of unwelded components. This is readily apparent from Fig. 17. The unwelded details in the figure will be familiar from Fig. 15, but in this case the test data have been obtained from high tensile structural C-Mn steel. Whereas the endurance limit for an unnotched steel component may correspond to a stress range of around 0.5 × ultimate tensile strength, so that fatigue failure would never occur at lower stresses, welded joints in steel can fail by fatigue at stress ranges below 30 N/mm^2. Moreover, fatigue failures can occur in welded joints under nominally compressive stresses. There are particular features of welded joints which produce these characteristics, which strongly contrast the behaviour of welded and unwelded material, and they are considered in this chapter in some detail. They are not the only factors which

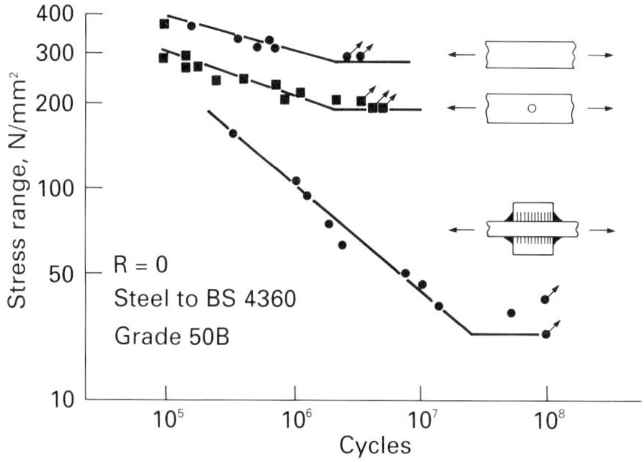

17 Comparison between fatigue strengths of plain steel plate, notched plate and plate with fillet welded attachments.

affect fatigue, others are discussed in Chapter 3, but they are so significant that they also affect the influence of these other factors.

Since most information about fatigue of welded joints refers to steel, the following discussion relates particularly to this material; later, fatigue strengths for joints in steel are quoted. However, much of the discussion would apply to welded joints in any material. Where significant differences arise, particularly with respect to aluminium alloys, they are noted.

Stress concentrations caused by weld shape

A weld bead made across a piece of plate would produce a change of shape and hence, if loaded in the direction shown in Fig. 18, a stress concentration similar to those described above. This concentration occurs at the 'toes' of the weld, that is at the junction of the plate surface and weld metal, and the magnitude of the peak stress will depend upon several factors. If there is a smooth transition between the plate and the weld metal the stress concentration will be low, but if, as is more likely under normal welding conditions, there is abrupt change of section, the stress concentration will be high. The geometry pro-

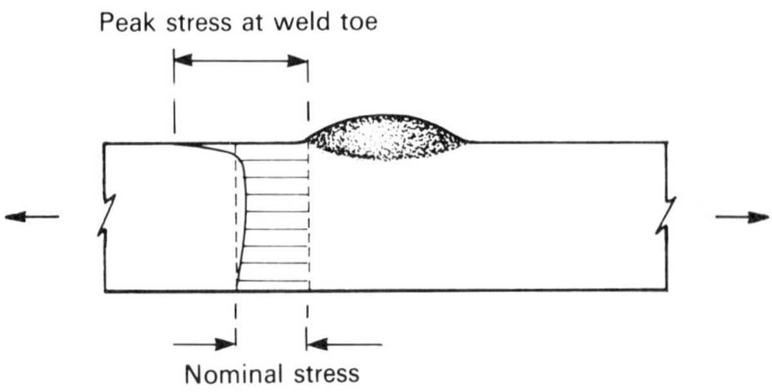

18 Stress concentration at toe of weld bead.

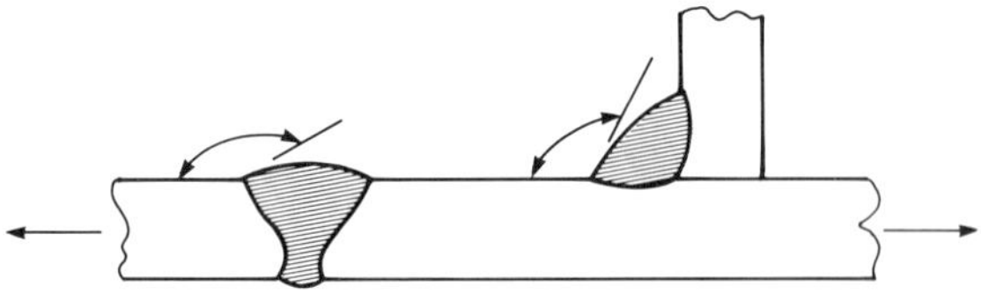

19 Comparison of butt and fillet weld shapes.

duced by the weld bead is similar to that produced in the transverse butt weld; in general, a transverse fillet weld produces an even more severe stress concentration because the change of section is more abrupt, Fig. 19.

It is to be noted that the weld itself need not carry or transmit the applied load to introduce a stress concentration. By its nature a weld forms an integral part of the member on which it is deposited, and, if the member is stressed, attachment welds at stiffeners, cleats, brackets and clips, Fig. 20, will produce severe stress concentrations by virtue of the sudden change of section. Fatigue cracks can propagate from these regions of stress concentration when the member is subjected to fluctuating loading; in other words, damage to a principal structural member might easily be caused by the unthinking or casual use of welding.

The weld toe proves to be the most likely site for fatigue cracking in welded joints loaded transverse to the toe. In continuous welds loaded parallel to the weld, the surface irregularities which cause stress concentrations include weld ripples and lumps due to stops/starts in the welding, Fig. 21, but these are less severe than the toe. Consequently, their fatigue strengths are higher than those of transversely loaded joints.

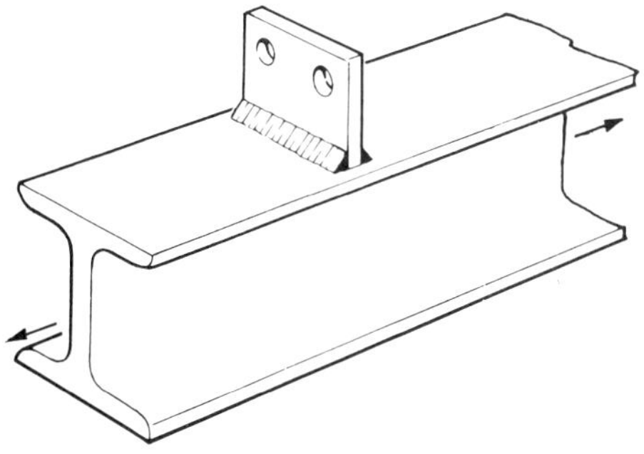

20 Source of stress concentration in loaded member because of non-load-carrying welded attachment.

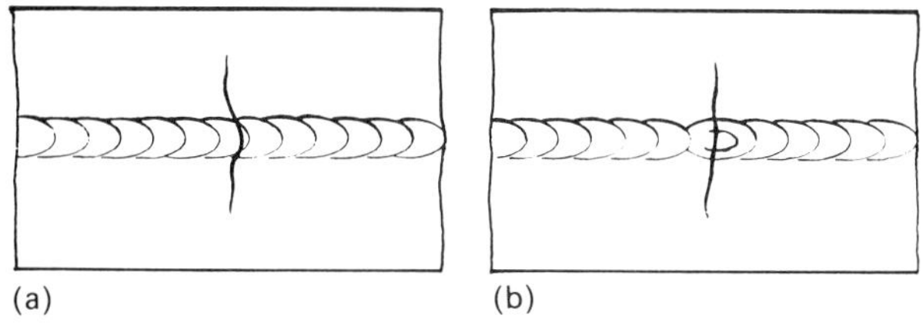

21 Fatigue cracking from: (a) Weld ripples; (b) Stop/start positions under loading parallel to weld.

Stress concentrations caused by discontinuities

Weld toe conditions

From stress analysis of an idealised model representing the cross-section of a fillet welded joint it can be shown that the stress concentration factor, K_t, at the weld toe under loading transverse to the weld is about the same as that at the edge of a hole. However, the fatigue strength of an actual weld is much less than that of the plate with a hole in it, as illustrated in Fig. 17. The difference arises because the weld toe geometry is more complex than the general profile suggests. Undercutting of the plate surface often occurs at the toes of welds and the weld may adopt an abrupt, convex profile, Fig. 22. Both these features will further increase the local stress concentrations. However, more important is the finding that very small crack-like discontinuities, termed 'intrusions', exist at the weld toe, Fig. 23, these being a product of conditions during welding which arise with most of the arc processes.

The weld profile and the extent of any undercut may vary along the length of the weld, particularly with manual welding processes, and

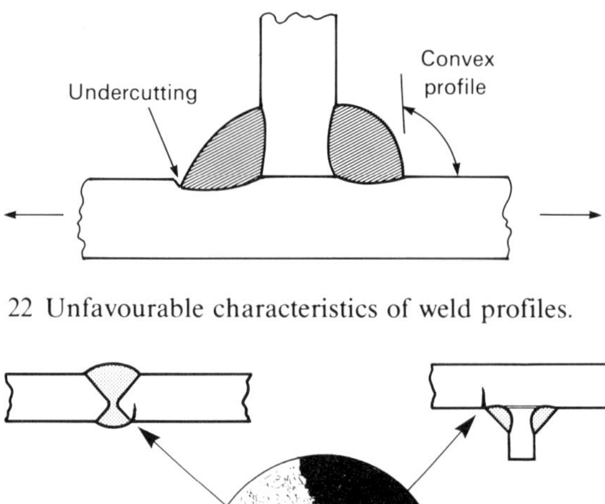

22 Unfavourable characteristics of weld profiles.

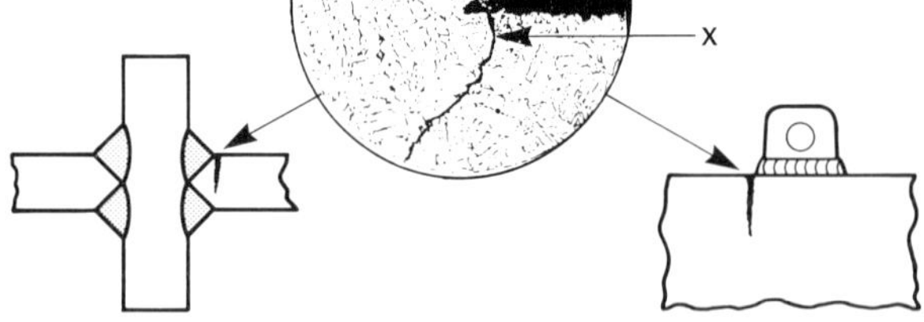

23 Fatigue cracking from pre-existing crack-like flaw (intrusion) which is an inherent feature of weld toes in steels (× 210).

the depth of intrusions also varies from one weld to the next. Clearly, in such circumstances the estimation of K_t values by calculation or experimental techniques becomes, to a large extent, meaningless for welded joints.

Stress concentration conditions at weld toes are thus compounded by the existence of the crack-like intrusion within a region of concentrated stress produced by the general weld profile and, possibly, undercutting. The result is that fatigue strength is reduced below that anticipated on the basis of the nominal weld profile, mainly because the number of stress cycles needed to initiate a fatigue crack is greatly reduced. Indeed, a striking contrast between welded and unwelded details is that, in the former, fatigue crack initiation may occupy a very small proportion of the total fatigue life, whereas in the latter it may occupy most of it. Looked at simply in terms of plates of the same steel and thickness, one as-rolled and the other with a fillet welded attachment, the effect of the difference is readily apparent from the corresponding S–N curves for repeated tension loading, R=0, Fig. 24.

Although weld toe intrusions have been observed only in steel, there is evidence to show that fatigue cracks also initiate very early in the lives of fillet welds in aluminium alloys.

Variations in the condition of the weld toe are always present and their effect on fatigue strength is, as far as possible, incorporated into S–N data for the joint, so that the designer does not need to consider them further. It is essential, however, that welds are made in accordance with procedures which avoid the incidence of weld toe flaws such as heat affected zone, hydrogen-induced cold cracking, and, for manual processes, with a sufficient degree of skill to produce welds free from deep weld toe undercutting.

Weld root condition

Partial penetration welded joints, Fig. 25, contain another feature which, under transverse loading, causes local stress concentration, namely the weld root. Depending on the joint geometry and extent of

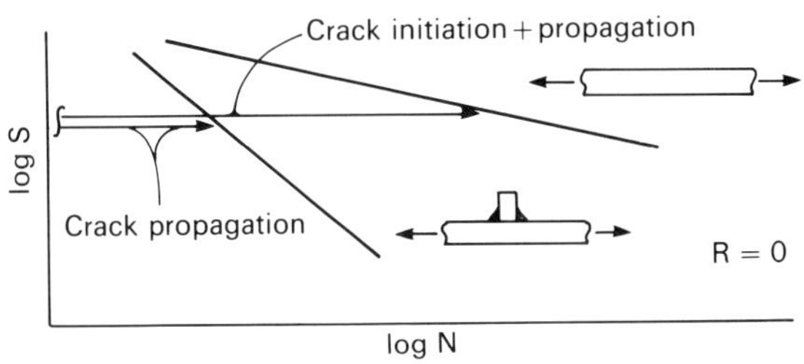

24 Schematic illustration of difference between S–N curves for plain (unnotched) plate and welded joint.

weld penetration, this can be more severe than the weld toe and become the site for fatigue crack initiation. The crack then propagates across the weld throat, Fig. 25. As in the case of weld toe intrusions, the lack of penetration zone is crack-like and therefore the fatigue life for this mode of failure also consists mainly of the propagation of fatigue cracks.

Unlike weld toe intrusions, incomplete penetration must be considered by the designer, either as an intended design feature (e.g. as in a cruciform fillet welded joint) or as an unintended welding flaw (see Chapter 3).

Residual stress

Operations such as casting, forging and rolling invariably produce 'locked-in' stresses in the product concerned and welding operations also induce stresses of this kind, the existence of which is independent of external loading. It is a natural condition that locked-in stresses are balanced within the body of material; in other words there is a system of tensile and compressive components of stress which is in equilibrium.

Two systems of stress are produced in the welded structure. Firstly, a structure will contain an overall distribution of stresses arising from assembly conditions and generally affecting the various members as a whole; such stresses are known as **reaction** stresses (or long-range residual stresses) and, to take the simplest concept, equilibrium of the reaction system could be achieved because tension in some members is balanced by compression in others. Additionally, **residual** stresses occur at each joint in a structure in a manner which makes it possible, if not strictly accurate, to consider them as localised stresses which might, therefore, influence the behaviour of the joint.

In outline, residual stresses arise as a consequence of the fact that, during the weld heating and cooling cycle, the natural expansion and contraction of material close to the heat source is inhibited by the

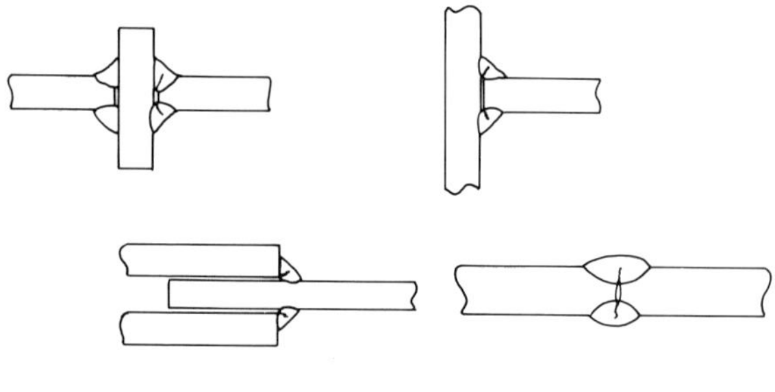

25 Fatigue cracking from the root of transversely loaded fillet or partial penetration welds.

restraining effect of adjacent material at a lower temperature. For example, during the cooling stage the longitudinal shrinkage of the weld metal is resisted, Fig. 26, and it may be visualised that the weld has had to accommodate its 'unnatural' length by plastic strain and, when cold, is subjected to longitudinal tensile stress. In fact, a stress of yield point magnitude will exist.

The actual situation is an as-welded joint (one in which residual stresses have not been relieved by postweld treatment) is complicated by practical factors such as the type and size of joint, the welding process used, and the way in which the weld metal is deposited. Nevertheless, the principle of high residual tensile stress in the weld metal (and adjacent parent metal) remains valid and can be characterised by considering idealised stress distributions for a simple butt joint, Fig. 27.

The development of longitudinal stress was indicated by the conditions shown in Fig. 26 and it follows that, for equilibrium, tensile stress in the weld zone, A, Fig. 27a, reaching yield point value, is counterbalanced by compressive stresses farther out in the plate, areas B_1, B_2. If uniformity of stress through the thickness is assumed (variations do occur in practice), equilibrium requires that area A equals areas B_1+B_2. Stresses acting transversely to the joint are also developed. The distribution of these is often less regular than the pattern shown, Fig. 27(b), but does include areas of high tensile stress coincident with the weld zone and ultimately complies with the same end result that transverse stresses must also be in equilibrium, as indicated.

Consider now what happens to the longitudinal residual stresses

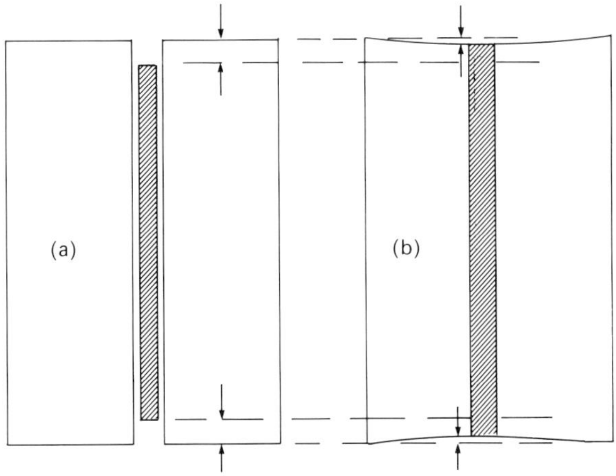

26 Formation of residual stress as a result of welding: (a) Natural longitudinal shrinkage of unrestrained (unattached) weld; (b) Longitudinal shrinkage of restrained weld.

when the plate is subjected to a nominal tensile stress, S_{max}, applied in the direction parallel to the weld. As the weld metal is already at the yield point of the material, S_{YS}, further local plastic straining occurs and the stress in the weld remains at S_{YS}, while the surrounding stress field changes to accommodate the applied load. This change is shown by the dashed line in Fig. 28a. If the applied stress is now removed, the residual stress distribution is altered to that shown by the chain-dotted line and the stress at the weld becomes $S_{YS}-S_{max}$. It follows, therefore, that if S_{max} is repeatedly applied, i.e. fatigue loading ranging from zero to S_{max}, the actual stress at the weld will cycle between a minimum value of $S_{YS}-S_{max}$ and a maximum value of S_{YS}, i.e. a range equal to the nominal range applied, but with a mean value different from that calculated from $S_{max}/2$.

If, instead of a tensile stress, a compressive stress, $-S_{max}$, is applied to the plate in the same direction, no plastic straining of the weld metal occurs, and the stress at the weld is reduced from S_{YS} to $S_{YS}-S_{max}$ as shown by the dashed line in the second diagram, Fig. 28b. On the removal of the applied stress the initial residual stress system remains unaltered. Under repeated compression fatigue loading, zero to $-S_{max}$,

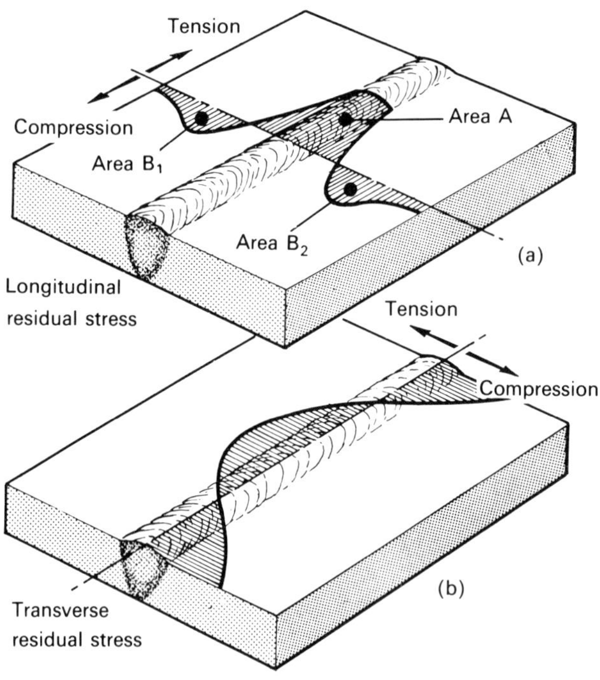

27 Typical residual stress distribution in welded joint.

the actual stress at the weld cycles from $S_{YS}-S_{max}$ to S_{YS}. Thus, for the same range of applied stress, the stress conditions produced at the weld are the same for both tensile and compressive applied loading, Fig. 29.

This is a simplified analysis which ignores, for example, the effect of the stress concentrations associated with the weld. However, it highlights two important consequences of the presence of tensile residual stresses of yield point magnitude. Firstly, fatigue failure can occur under loading conditions which, nominally, introduce compressive stresses; secondly, the fatigue strength of welded joints is governed by the applied **stress range** regardless of the nominal applied stress ratio, at least for R values down to $R=\infty$. Therefore, fatigue design is based on stress range and a single S–N curve represents the performance of a given welded joint for any minimum/maximum ratio of load input, as indicated in Fig. 30.

The magnitude of the residual stresses introduced by welding depends on the tensile strengths of the weld and parent metal. In steels

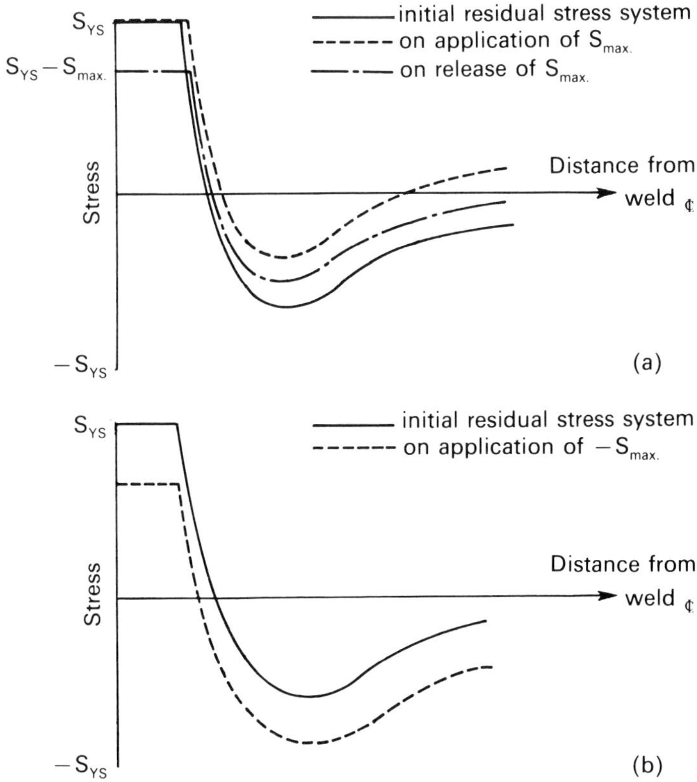

28 Superposition of applied stress and residual stress of tensile yield magnitude.

the two are usually closely matched but in heat-treatable aluminium alloys the as-deposited weld can have a lower strength than the parent metal, and consequently residual stresses are not as high as the yield strength of the parent metal. This means that applied stress ratio may still influence fatigue strength; however, failures can still occur under nominally compressive stresses.

Although it must be recognised that fatigue cracks in welded structures under nominally compressive stresses can, and do, occur, it may be noted that they will propagate only in a region of tensile residual stress. The formation of the crack will redistribute the residual stresses and this may provide some relief of the tensile residual stress. In such a

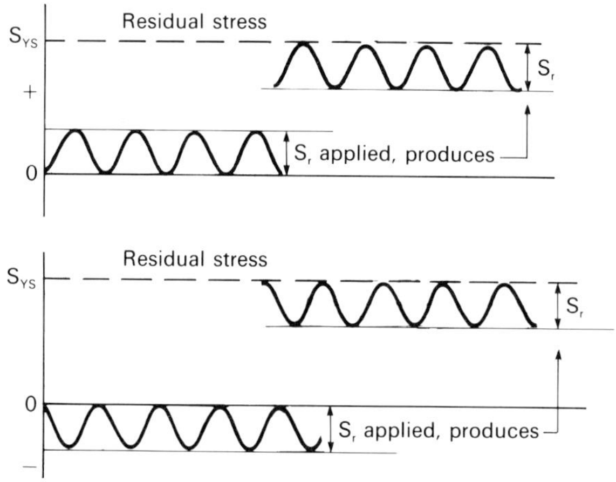

29 Effective stress resulting from superposition of applied and residual stress.

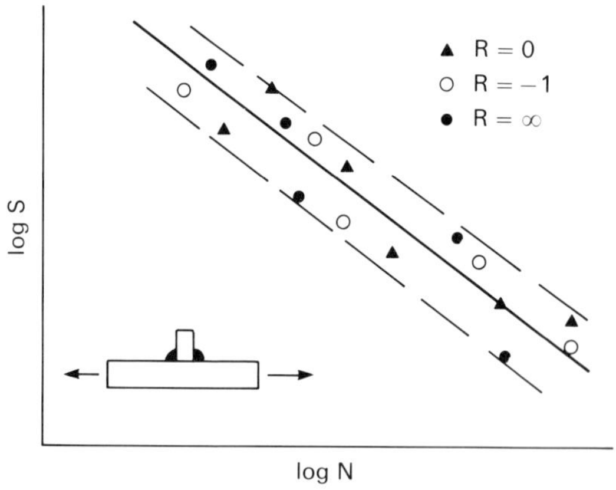

30 Fatigue test results for steel fillet welded joints containing high tensile residual stresses.

case, the actual stress at the crack may revert to the compressive stress system being applied and the crack no longer propagates. The consequence of this type of cracking in any particular structure can be assessed only by the designer. In a highly redundant structure it may not be important, but in other situations the extent of fatigue cracking may still be critical (e.g. leakage in a container) or sufficient to accelerate corrosion. Also, in any case which involves both applied tensile and compressive stresses, the fatigue damage contribution due to the compressive stresses must always be taken into account.

Just as tensile residual stresses can lead to a reduction in fatigue strength so compressive residual stresses, which also exist in the as-welded structure, are beneficial. However, because such compressive stresses are accompanied by tensile stresses this benefit cannot be used in design. There are methods for the introduction of local compressive residual stresses after welding which can be used to increase the fatigue strength of welded joints by being able to control the position of the favourable compressive stress field, but this aspect is not pursued here (see Pt 2).

Practical implications

The worst combination of the features described results in welded joints containing yield magnitude tensile residual stresses having geometries which introduce severe stress concentrations combined with crack-like discontinuities, from which fatigue cracks readily initiate under either tensile or compressive applied repeated stresses. Compared with unwelded details, the resulting joint has a relatively low fatigue strength, a steeper S–N curve and a dramatically reduced endurance limit, as illustrated in Fig. 17. In practice, the fatigue behaviour of welded joints varies, depending on the type of the joint and the direction of loading, but the above combination of features is far from uncommon and certainly the geometric conditions, which are significant for transverse applied loading, arise in virtually any fillet or butt.

The lack of a significant fatigue crack initiation period, due to the presence of crack-like discontinuities (e.g. weld toe intrusions, incomplete weld root penetration) is particularly significant in that it has other practical implications. The factors which govern the fatigue crack initiation process, which may occupy most of the life in an unwelded component, can have quite different effects on fatigue crack propagation, which occupies most of the life of the welded joint. As a result, there are important contrasts between the fatigue behaviour of welded and unwelded material and conclusions based on the fatigue behaviour of the material can be quite misleading when applied to welds. This will be apparent in the discussions in Chapter 3.

Chapter 3

Factors which affect fatigue of welded joints

Stress relief

The presence of high tensile residual stresses in welded joints is clearly undesirable and there would seem to be some incentive for their removal by stress relief. However, the benefit to be obtained by this treatment may not justify the cost involved. Comparing the fatigue behaviour of as-welded joints containing high tensile residual stresses with those which have been thermally stress relieved, tested over a range of stress ratios, it is found, Fig. 31b, that the residual stresses are important only if the loading produces partly or wholly compressive stresses; and then only in the high-cycle fatigue regime. For fully tensile applied stresses the results for as-welded and stress relieved joints are virtually the same, the small difference amounting to around 15% on stress range at 2 × 10⁶ cycles (Fig. 31c). This is a consequence of the

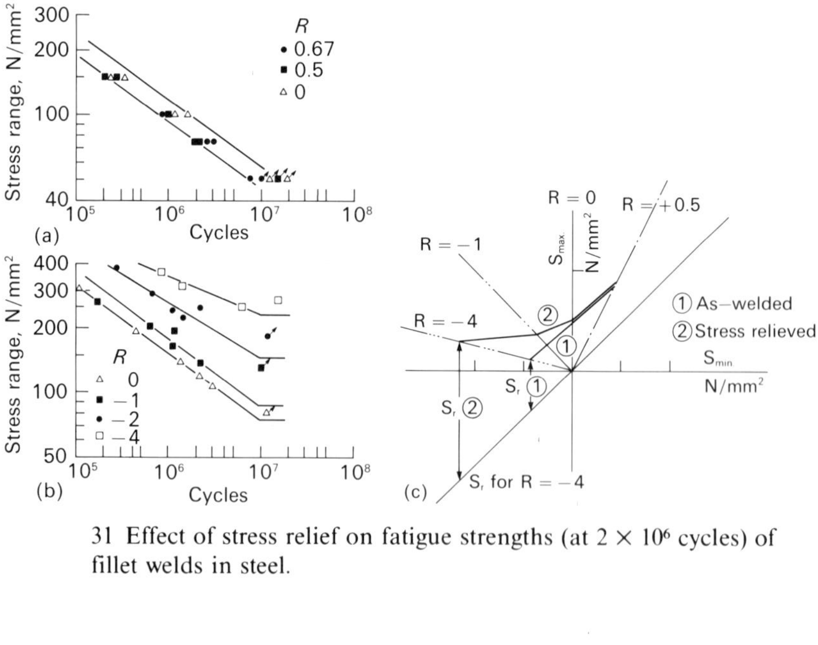

31 Effect of stress relief on fatigue strengths (at 2 × 10⁶ cycles) of fillet welds in steel.

dominance of fatigue crack propagation and the fact that, in weldable steels, rate of crack growth is hardly influenced by mean stress for fully tensile conditions. However, as the applied stress cycle becomes more compressive, the stress range needed to produce failure in a given life increases. This is because the compressive part of the applied stress cycle is largely non-damaging in stress relieved joints so that part of the cycle must remain tensile to produce failure. Thus, it will be apparent that stress relief of welded joints is of significant benefit only if the applied loading introduces stress fluctuations partly or wholly in compression.

Fatigue crack growth in aluminium alloys is more sensitive to mean stress. As a result, S–N data for welded joints with low residual stresses may exhibit a mean stress effect for compressive and tensile applied stresses. However, the effect will still be strongest when the stress is partly compressive.

Material properties

Perhaps the most striking contrast between the welded and unwelded material, which comes about as a result of the differing significance of the crack initiation process, is the influence of material strength. The fatigue strengths of unwelded components normally increase with the material tensile strength. However, this is not the case with welded joints in which fatigue failure is by crack propagation from crack-life flaws, as illustrated in Fig. 32. This behaviour is consistent with the fact that rate of crack growth does not vary significantly, and certainly not consistently, with material tensile strength. Thus, any beneficial effect of tensile strength on fatigue must relate to the crack initiation process,

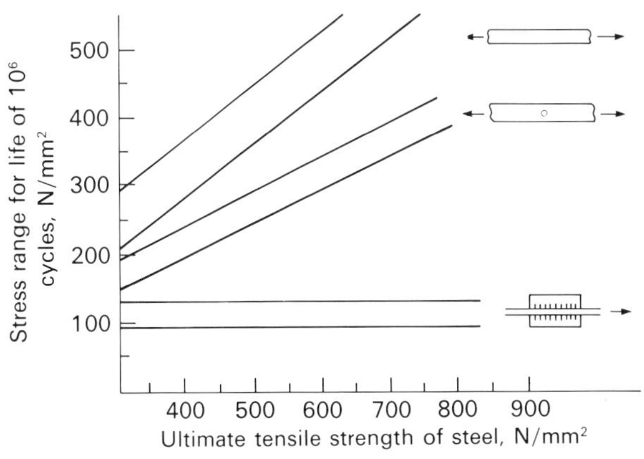

32 Effect of steel strength on fatigue strength.

which may form a significant proportion of the fatigue lives of unwelded parts.

Although fatigue cracks may propagate through relatively complex microstructures associated with weld metal, diluted weld metal and heat affected zone, this rarely has an influence on fatigue strength. This is because rate of crack growth is relatively insensitive to microstructure, apart from exceptional circumstances when it allows a change in fracture characteristics such that the normal fatigue process (striation formation) is accompanied by a static mode of fracture (e.g. cleavage).

The presence of brittle material might also reduce fatigue life as a result of premature fracture. However, in most practical cases fatigue crack growth rate increases exponentially and hence, unless the critical crack depth is extremely small, the crack size at final failure does not have a large influence on fatigue life.

Another consequence of the dominance of fatigue crack propagation in the lives of welded joints is that reasonable correlation of fatigue data for joints in different materials can be achieved on the basis of strain range. Thus, the fatigue strength of a fillet weld in aluminium alloy is approximately ⅓ of that for the same joint in steel, that is their fatigue strengths are in the ratio of their elastic moduli.

Weld quality

Apart from the highly significant inherent flaws at the toes of welds, other imperfections can occur in welded joints and they may influence the fatigue life. These include welding flaws such as porosity, slag inclusions, lack of fusion and incomplete weld root penetration. Such flaws provide additional stress concentrations which, if they are more severe than those already arising from the basic weld geometry, will provide alternative sites for fatigue crack initiation and could therefore lead to a reduction in fatigue life. Clearly, this is most likely to be the case in joints with relatively high fatigue strengths. This is borne out by the fact that welding flaws are of greatest significance in flush-ground butt welds and continuous welds stressed parallel to the weld, that is cases where the weld toe is either absent or parallel to the loading.

Another type of imperfection is misalignment, including axial misalignment due to eccentricity between abutting plates and angular distortion, such as peaking in pipe or vessel seam welds, Fig. 33. The effect of misalignment is to increase the stress experienced by the joint when it is loaded, as a result of the introduction of local secondary bending stress, and hence reduce the fatigue life. As in the case of stress concentrations due to weld geometry, misalignment is only harmful if it presents a discontinuity to the applied load path. Thus, secondary bending stresses do not arise as a result of misalignment in continuous welds loaded longitudinally (i.e. parallel to the discontinuity). Also, misalignment has no effect on fatigue strength if the only loading on the joint is bending. The distinction between misalignment and the welding flaws mentioned earlier is that it does not introduce an alternative site for fatigue crack initiation but rather it enhances the severity of existing stress concentrations, notably the weld toe since the secondary bending stress is highest on the plate surface.

Apart from the potential problem of misalignment as a feature which could reduce the fatigue life of an actual welded structure, its presence in laboratory test specimens causes problems in the interpretation of fatigue test results, as discussed in more detail in Chapter 4.

The use of correct welding procedures, including the control of joint preparation and fit-up, helps to reduce the occurrence of welding imperfections. Nevertheless, it is sometimes necessary to assess whether or not imperfections revealed during inspection of a welded joint are acceptable. A rational basis for the decision is fitness-for-purpose, whereby an imperfection would be considered acceptable provided its presence did not reduce the fatigue life of the joint below that required for the intended service. In this respect, it is important to recognise that some flaws occurring in practice do not reduce the fatigue life (see Fig. 44) and indeed even if the life was reduced, it may still be sufficient to achieve the design lifetime.

Application of the fitness-for-purpose concept requires knowledge of the stress concentration effect of welding flaws and the fatigue

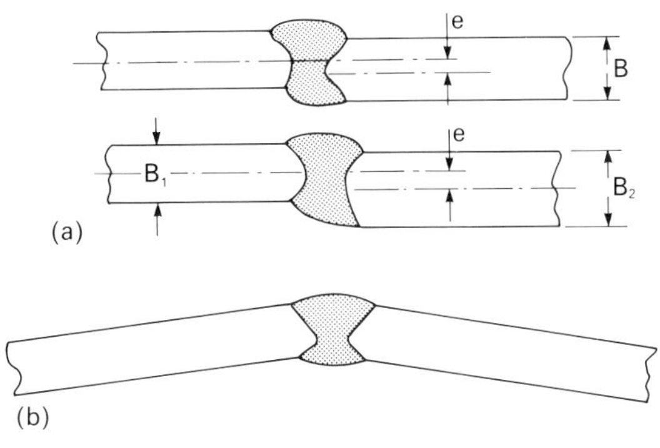

33 Misalignment in butt welded joints: (a) Axial; (b) Angular.

performance of welded joints containing imperfections. Such information is available and has been used to develop flaw acceptance limits, as discussed further in Pt 2.

Size effects

A size effect in the fatigue behaviour of machine components, whereby fatigue strength decreases with increase in size, is well known. It can be explained statistically in that the number and severity of flaws is likely

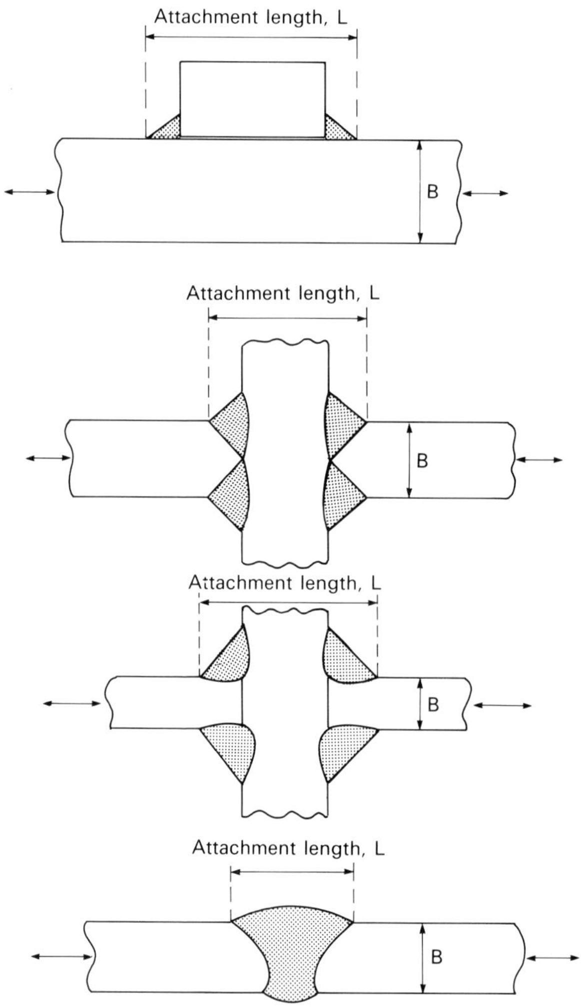

34 Dimensions relevant to size effects in transverse fillet and butt welded joints.

to increase with size, and also in terms of the effect of relative dimensions on any geometric stress concentration present.

A size effect also arises in members subjected to bending, in that fatigue strength increases with increase in stress gradient. Thus, a high stress gradient in a thin plate in bending gives rise to a higher fatigue strength than that of a thick plate.

Similar size effects are found in welded joints, but in this case the statistical explanation is probably the least significant. Of particular

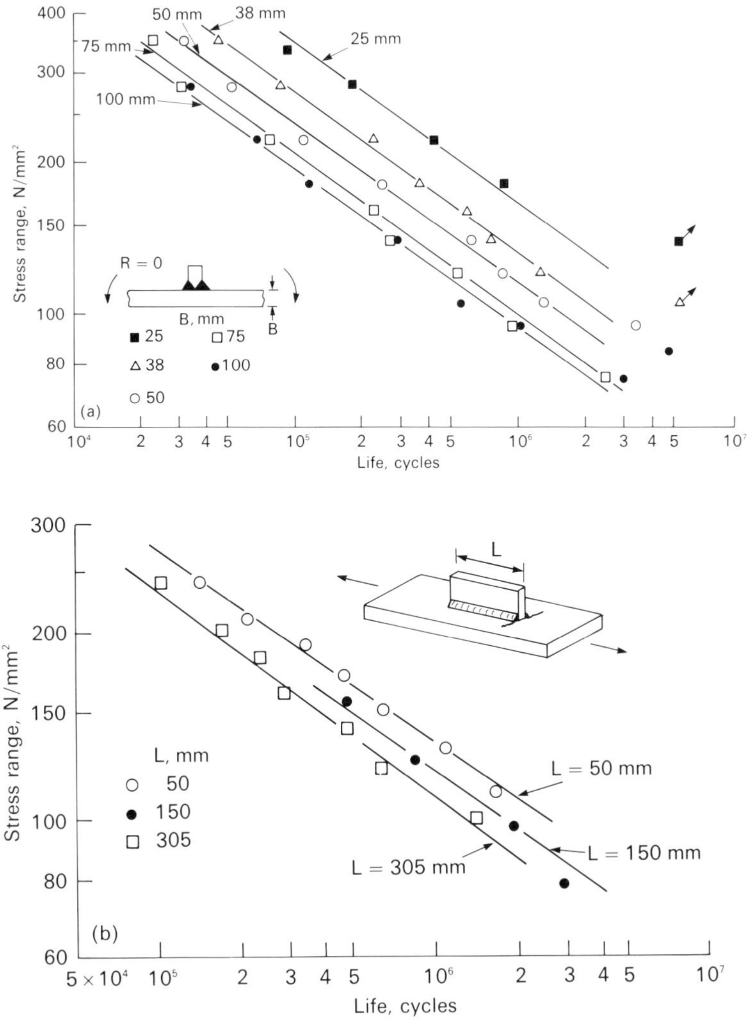

35 Fatigue test results showing influence of: (a) Plate thickness; (b) Attachment length.

importance is the influence of welded joint dimensions on stress concentration. The effect is particularly pronounced when failure of the welded joint consists of fatigue crack propagation from a surface stress concentration, such as the weld toe. Considering the simple case of a stressed plate with a fillet welded attachment, Fig. 34, it is found that the magnitude and extent of the stress concentration at the weld toe increases, and hence fatigue strength decreases, with increase in attachment length (L) and stressed plate thickness (B). Examples of the influence of these dimensions on the fatigue behaviour of fillet welds are given in Fig. 35. A similar situation arises with transverse butt welds where weld bead width is analogous to attachment length (see Fig. 34).

The width of the stressed member might also be significant, as in the case shown in Fig. 36.

It is likely that, in future, methods will be developed for incorporating the effect of welded joint dimensions when calculating the stress experienced by the joint. However, at present most fatigue design is based on S–N curves expressed in terms of the nominal stress experienced by the joint, with all stress concentration effects associated with the joint incorporated in the test data. Thus, in such cases it is necessary to ensure that the design data refer to the particular dimensions of the structure being designed, or to introduce appropriate correction factors to data relevant to other dimensions, as discussed more fully in Pt 2. Fatigue strengths quoted in this book refer to fatigue test results obtained mainly from specimens 12 to 25 mm thick, the vast majority at the lower end of this range.

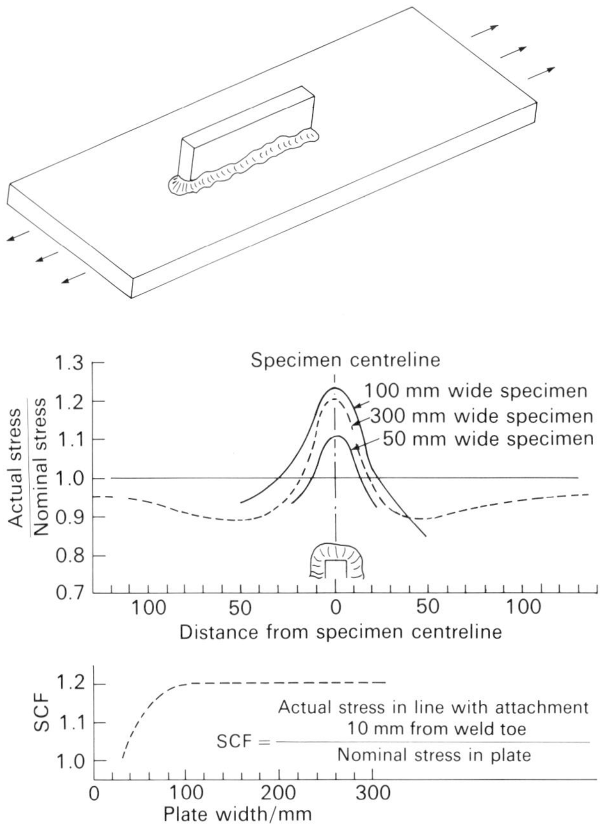

36 Effect of plate width on stress concentration factor (SCF) at end of longitudinal attachment. (NB–SCF does not include the effect of the weld itself.)

Chapter 4

Fatigue of welded joints

Transverse butt welds

A butt weld is perhaps the most efficient way of joining two pieces of plate. A full penetration transverse butt weld, i.e. one which lies at right angles to the direction of loading, between two pieces of steel, tested under conditions giving static ductile failure, would produce a 'full strength' joint, i.e. failure would normally occur in the parent material away from the weld with no reduction in strength caused by the weld. However, if the joint is to be subjected to fatigue loading, the description 'full strength' is a misnomer because the stress concentration associated with the weld will reduce the fatigue strength of the joint compared with that of the parent plate. Referring to a previous diagram, Fig. 15, it will be seen that the fatigue strength of mild steel plate in the as-rolled condition is 250 N/mm^2 at 2×10^6 cycles of repeated tension loading. This compares with 170 N/mm^2 for the plate with a hole and 95 N/mm^2 for the notched plate. Similar tests on a series of specimens containing nominally good quality transverse butt welds made with the manual metal arc (MMA) process would give a fatigue strength of about 155 N/mm^2 at 2×10^6 cycles, representing a fatigue strength reduction of 38% compared with the unwelded parent plate.

Fatigue failure of the joint considered consists of crack initiation at one or more weld toes and crack propagation through the thickness of the plate normal to the direction of loading, Fig. 37. It does not occur as a result of unsound welding or some particular properties of the weld metal but simply as a consequence of the presence of the stress concentration at the weld toe, including the effect of the weld toe intrusions (see Fig. 23). As would be expected, complete removal of the excess weld metal and weld toe intrusions by machining or grinding can restore the strength of the joint to a level where fatigue failure would then normally occur in the parent plate. In such circumstances weld flaws may, however, be of great significance.

It is possible, therefore, to achieve higher fatigue strengths from transverse butt welds than 155 N/mm^2 at 2×10^6 cycles, but much lower strengths might also be obtained. Differences arise because variations occur in the shape of the weld profile, the extent of undercut and the alignment of the joint which, as noted earlier, lead to marked changes in the stress concentration at the weld toe. The fatigue strengths of transverse butt welds are particularly widely scattered, varying from 230 N/mm^2 to as little as 80 N/mm^2 at 2×10^6 cycles.

This wide range of strengths is not dependent on the presence of internal defects, and from the point of view of static strength each weld form would be considered acceptable. Clearly, for fatigue loading conditions the term 'weld quality' must also refer to the profile of the excess weld metal and the alignment of the joint.

The mistaken idea still persists that the provision of generous excess weld metal is one way of making sure of adequate joint strength. This gives rise to the use of the misleading term 'weld reinforcement' which could be better described as 'overfill'. It can be seen from the macrosections, Fig. 38, that the lower fatigue strength is associated with the weld having a poor profile, in that there is a very sudden change of section at the junction between parent plate and weld metal. Welds which have a minimum of excess metal and a smooth transition at the weld toe give the highest fatigue strengths. In general, the MMA process is the most reliable for the production of these features. Although they can also be secured by the use of an automatic welding process, it is a consideration that the very worst butt weld shapes can also be produced by such processes, particularly submerged-arc welding, if the welding parameters are not carefully and correctly controlled. Unless very elaborate quality control requirements are introduced, the designer has little control over the shape of the butt welds made by a given welding process which will be incorporated in a structure. It is

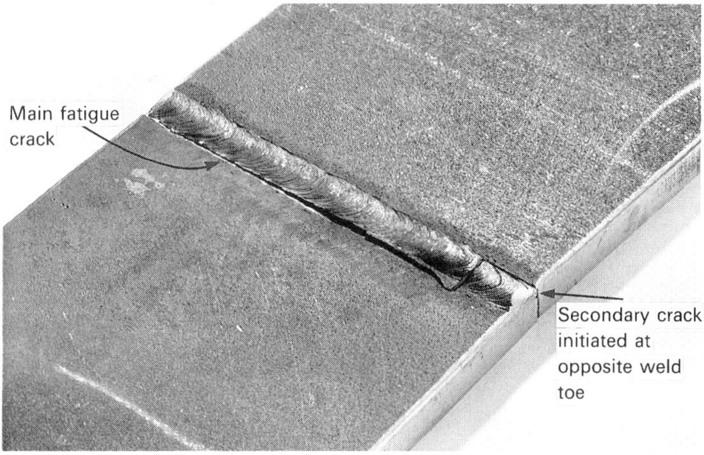

37 Fatigue failure from the toe of a transverse butt weld.

important, nevertheless, to ensure that the welder is given the best possible chance of making a weld of desirable form. Ease of access, good fit-up, correct joint preparation, and welding in the flat position are all factors which will improve this chance. Whether the welds are to be made in the fabrication shop or on site is also an important influence in their quality.

Whilst there is clear evidence for an influence of weld profile on the fatigue strengths of transverse butt welds, the full significance of that influence is confused by another aspect of weld quality to which transverse butt welds are particularly susceptible, namely misalignment. As noted earlier, the effect of misalignment is to increase the stress at the weld toe as a result of the introduction of secondary bending. Thus, expressing the fatigue strength of a misaligned test specimen in terms of the usual nominal stress range in the plate will underestimate that stress, and variations in misalignment from one specimen to another will contribute to scatter in the fatigue test results. The significance of misalignment will be readily apparent from estimates of the magnitude of the stress induced. For example, in the simple case of axial misalignment e (see Fig. 33) in a butt weld between plates of thickness B, the local bending stress $S_b = \frac{3e}{B} \times S_a$, where S_a is the nominal applied axial stress. Thus, the total stress at the weld toe in the plate is $S_a + S_b = S_a\,(1 + 3\frac{e}{B})$, the term in brackets being the stress magnification factor, termed K_m. Thus, for axial misalignment of only 10% of the plate thickness (i.e. $\frac{e}{B} = 0.1$), $K_m = 1.3$. A similar increase in stress can arise

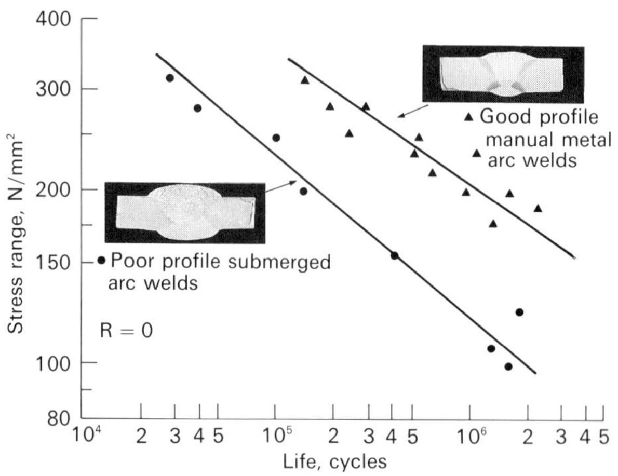

38 Effect of weld profile on fatigue strength of transverse butt welds.

as a result of only 1° angular misalignment, while a combination of both these slight imperfections gives $K_m = 1 + 0.3 + 0.3 = 1.6$. The effect on fatigue life of misalignment in transverse butt welded joints in steel is illustrated in Fig. 39(a). Replotting the results in terms of the total stress range K_m S_a, Fig. 39 (b), confirms that the variation in fatigue strength seen in Fig. 39(a) was due entirely to the introduction of secondary bending stresses. This is particularly significant in relation to the argument about the relative importance of weld profile and misalignment. The specimens containing the greatest misalignment (100% of plate thickness) were effectively lap joints in which the weld profiles resembled fillet rather than butt welds. In spite of this 'poor profile', the fatigue strength was comparable with that obtained from good profile transverse butt welds.

In practice, it is often difficult to avoid slight misalignment, especially angular distortion, in butt welded test specimens. This is a particular problem if the specimens are sliced from a welded panel, as alignment

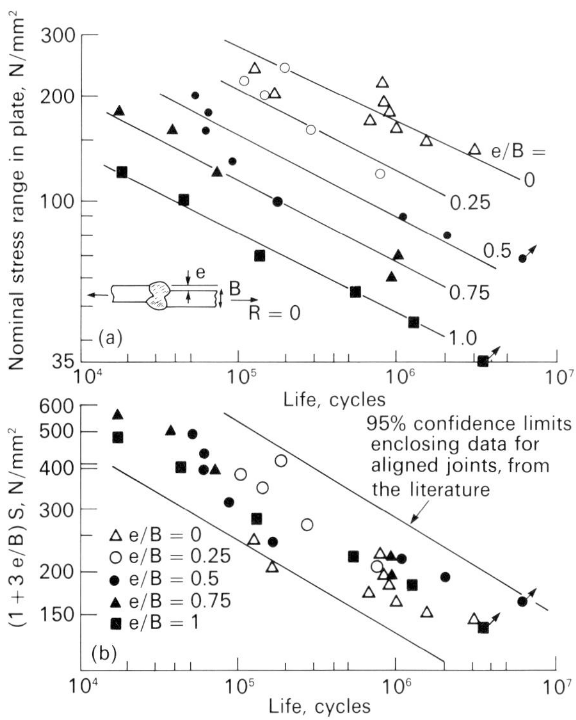

39 Fatigue test results from axially misaligned butt welds expressed in terms of: (a) Nominal stress range, S; (b) Local stress range $K_mS = (1 + 3e/B)\ S$.

and fit-up can change as the welding progresses. Difficulties can also arise when the weld is made using an automatic process, like submerged-arc welding, since it is more difficult to balance the welding to minimise distortion than in manual welding. The same will also be true of welds in real structures but then the effect of misalignment may be less pronounced. The simple test specimen is free to bend under the induced bending moment due to misalignment and hence experience the full harmful effect of the misalignment. However, joints in real structures are likely to be more restrained, by the surrounding structure or because of the nature of the structural section being welded. As a result, the introduction of secondary bending stresses will be inhibited. Thus, ideally it is desirable to obtain fatigue test data from perfectly aligned specimens and to take account of any misalignment in the real structure separately, as part of the design calculations.

Referring back to the discussion of the influence of weld profile and the fact that submerged-arc butt welds have given the poorest fatigue properties, it is significant that test specimens made using this process are the ones most likely to be misaligned. The practical implication of this possibility is that some of the published fatigue data for submerged-arc butt welds are misleadingly low, in that they incorporate the effect of some misalignment, and that weld profile shape may be less significant than was first thought to be the case.

The discussion so far has considered butt welds made from both sides in which failure initiates from one of the weld toes. In practice it is not always possible to gain access to both sides of a butt joint and all the welding may have to be done from one side only. Obvious and important examples are welds in pipes or rectangular or circular hollow structural sections.

What has been said in respect of the shape of the excess weld metal applies equally to the shape of the root or penetration bead. Welds having a good root bead shape with a smooth profile will have similar fatigue strengths to double-sided welds. Unfortunately the control of root penetration and profile is very difficult using normal welding techniques, and misalignment of parts, excessive penetration and root crevices can all easily occur. It will be appreciated that these features will produce lower fatigue strengths. The fatigue cracks will initiate at the edge of the root bead and propagate through the weld metal, Fig. 40. Fatigue failure resulting from such features can be a problem in circumferential butt welds in pipework subjected to repeated bending and in the brace members of tubular offshore structures.

Welding techniques have been developed which are intended to produce the required weld penetration and desired root bead shape. For example, the use of oxyacetylene or TIG welding for the first (root) run, the use of a fusible steel root insert also with TIG welding, or the use of special flux-coated or ceramic backing strips with certain welding processes.

Some success in producing high fatigue strength butt welds by welding from one side has also been achieved using the electron beam process. Indeed, the advantage that this process offers in terms of speed of welding could be lost if it was necessary to resort to welding from both sides. As would be anticipated from studies on arc welded joints, the fatigue strength is found to be highly dependent on weld root quality, the presence of lack of penetration or an unfavourable profile leading to a reduction in fatigue strength. However, if full penetration is achieved, electron beam welds have been shown to have fatigue strengths comparable with the best arc welds, even with relatively poor weld profiles, Fig. 41. This may be a further indication that weld profile

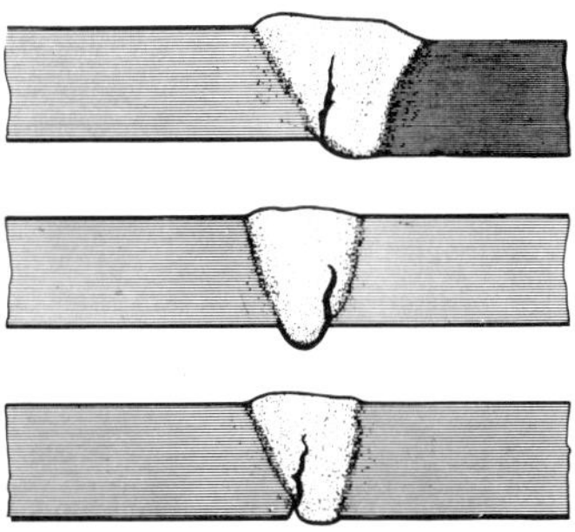

40 Unfavourable features which can arise at roots of butt welds made from one side.

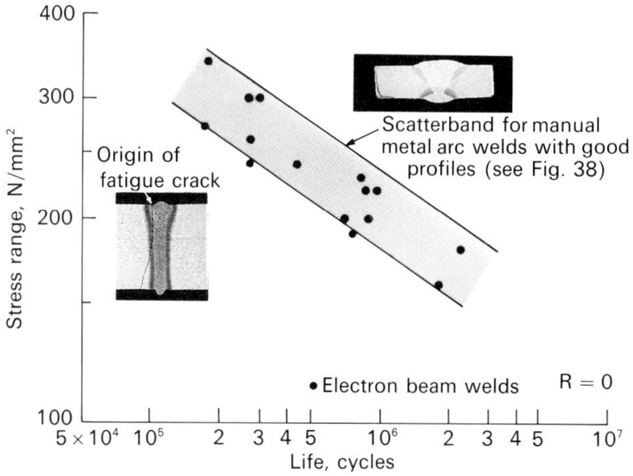

41 Fatigue strength of full penetration electron beam butt welds made from one side.

is not of prime importance, or that the narrowness of the weld bead results in a relatively low stress concentration factor in spite of the poor profile. Certainly, small 'attachments', analogous to the butt weld bead width, are known to result in lower stress concentration factors than large ones, as discussed under 'Size effects' in Chapter 3.

To obtain consistent results with the above techniques a high degree of control over the welding procedure is necessary, but this cannot always be justified. In such circumstances a permanent backing bar or strip is often used. Provided full fusion with the backing strip is achieved, fairly consistent fatigue strength can be obtained. In this event the critical stress concentration occurs at the junction of the weld metal and the backing strip. The fatigue crack will initiate at this point, Fig. 42, and propagate through the weld metal. The average fatigue strength of such a joint is about 115 N/mm^2 at 2×10^6 cycles. This is in the lower half of the range of values quoted for joints when they fail from the weld toe and is, in fact, similar to the fatigue strength of a fillet weld. However, this downgrading of the joint may be a small price to pay for a guarantee of full penetration welding.

The use of a permanent backing which is also integral with one of the members has the advantage that it assists with the location and setting up of the joint. This is particularly useful with tubular components, Fig. 43. In some applications of this kind a further improvement in fatigue strength can be obtained where it is possible to machine out the bore subsequent to welding.

As mentioned earlier, internal flaws in welds give rise to stress concentrations, the importance of which is dependent upon their severity relative to that of stress concentrations which stem from weld geometry. Two examples relating to the transverse butt weld illustrate this point. Figure 44a indicates a case in which a slag inclusion, embedded in the weld, did not cause loss of fatigue strength by precipitating cracking at an earlier stage than that associated with the weld toe. Small, distributed pores (porosity) in the weld metal can be similarly ineffective, Fig. 44b. However, both types of flaw would be significant if

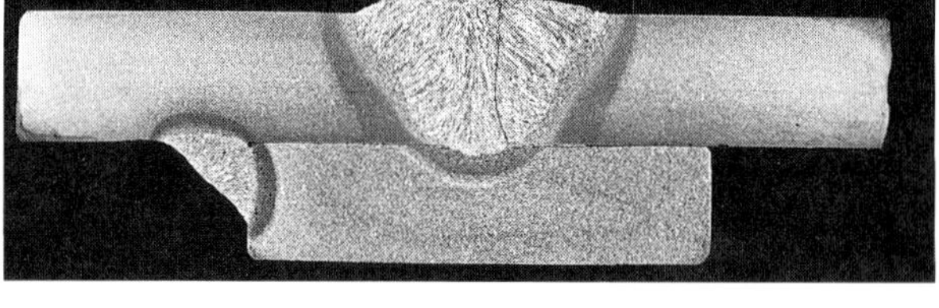

42 Fatigue failure in transverse butt weld made on backing strip (×2).

the weld was ground flush. On the other hand, incomplete penetration, representing a severe discontinuity normal to the direction of applied stress (in a transverse butt joint), can seriously affect fatigue strength, even when, as shown in Fig. 45, the weld profile is less than good. The comparative severity of stress concentrations thus forms the basis for the assessment of weld flaws, but note that, in turn, the severity of a planar flaw, such as incomplete penetration, is also determined by its orientation with respect to applied stress.

A final point regarding the fatigue behaviour of transverse butt welds concerns the ends of the weld if they occur at a free edge. As will be seen later, the stress concentration introduced by a weld is intensified if the weld is on a plate edge. This, together with the fact that the weld shape is likely to be poorer in a plate edge, means that the fatigue strength of the joint is likely to be reduced. To avoid this problem, butt weld ends should always be ground flush with the plate edge. In practice this presents little extra work since runon and runoff tabs, which should always be used in such circumstances, will need to be removed anyway.

Longitudinal butt welds

It has been seen that the fatigue behaviour of transverse butt welds depends primarily on the stress concentration at the toe of the excess weld metal. In longitudinal butt welds, the change of section caused by

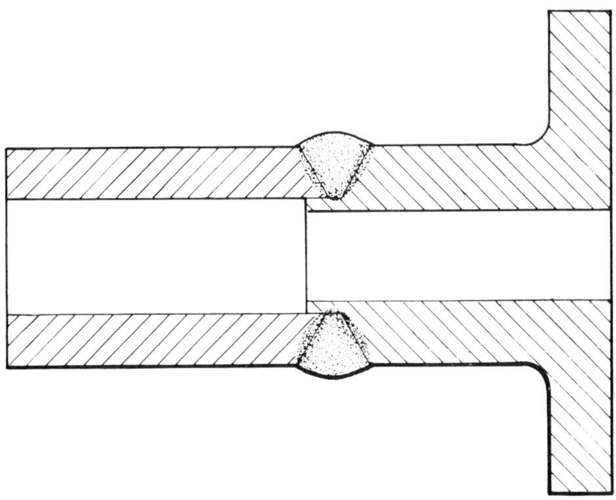

43 Flange designed to provide integral backing for tube-to-flange butt weld.

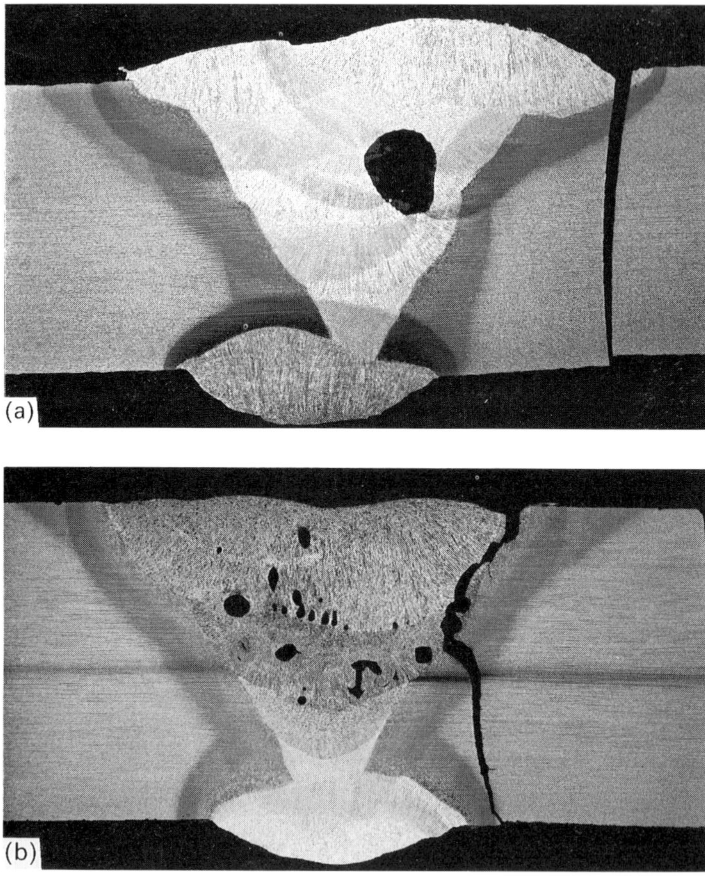

44 Butt welds containing: (a) Slag inclusion; (b) Porosity, which failed by fatigue crack growth from weld toe (×5).

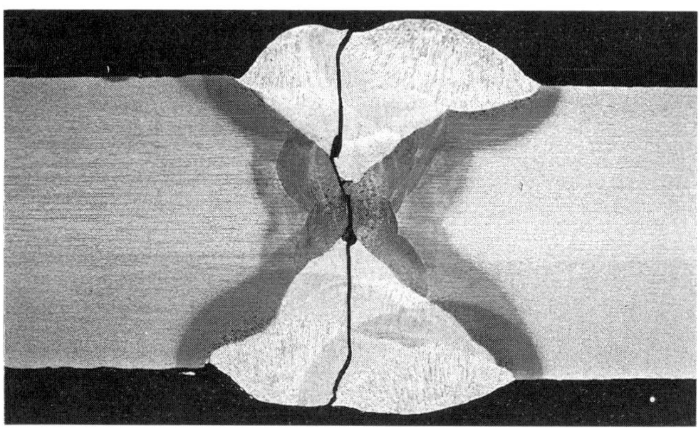

45 Butt weld containing lack of penetration defect from which fatigue cracking occurred (×5).

this excess weld metal lies parallel to the direction of applied load and therefore does not introduce a stress concentration. The features which act as stress concentrations, at which fatigue cracks initiate, are stop/start positions, for example where the electrode is changed in manual welds, and the ripples on the surface of the weld, Fig. 21. The stress concentrations associated with these features are far less severe than that at the edge of the excess weld metal in transverse joints, and consequently the fatigue strength of the longitudinal butt weld is higher: around 140 N/mm^2 at 2×10^6 cycles for MMA joints.

Welds made continuously with an automatic process will be free of stop/start positions and can be almost free from surface ripples. As a result their fatigue strengths tend to lie at the top of the range of strengths for the longitudinal joint.

The fatigue strengths of welds made from one side, or indeed welds made from both sides containing incomplete penetration, will no longer be sensitive to the transverse profile of the root bead or the weld flaw since these discontinuities lie parallel to the direction of loading. However, variations in weld penetration resulting in a rough root bead, Fig. 46, could be more significant than ripples on the weld surface. The use of a permanent backing strip would be beneficial in this situation since it would not introduce a stress concentration with respect to the longitudinal loading, provided it is continuous. When discontinuous backing strips are used it is found that hot cracking transverse to the weld invariably occurs at the weld root where the weld crosses the break. Not surprisingly, the resulting fatigue strength of the joint is very low, the actual strength depending on the extent of the cracking. In one test series the fatigue strength was only 48 N/mm^2 at 2×10^6 cycles. Clearly, discontinuous backing strips should never be used.

Even then, if a continuous backing strip is achieved by butt welding together shorter lengths, the butt welds must be fully penetrating, otherwise fatigue cracking may initiate from the lack of penetration because the backing strip becomes an integral part of the main member.

When continuous backing strips are used it is desirable to ensure that all tack welds holding the strip are so located that they are subsequently buried by the butt weld. The consequence of tack welding a

46 Variation in underbead shape in butt weld made from one side only.

backing strip in position with short lengths of external fillet weld, Fig. 47, would be to cause a reduction in the fatigue strength of the joint to 125 N/mm^2 at 2×10^6 cycles, as discussed below under 'Fillet welded connections'.

The good fatigue strength of longitudinal butt welds can be obtained only if they are continuous and the effect of weld ends can be avoided. The junction of the weld end and the parent plate is comparable with a weld toe in that crack-like intrusions, Fig. 23, are still an inherent feature. Consequently, a joint which has one of the lowest fatigue strengths of all welds is the butt weld between one plate and the edge of another which is under stress, Fig. 48. The example shown with butt welded side plates has a fatigue strength at 2×10^6 cycles of about 70 N/mm^2 based on the cross-sectional area of the stressed plate, i.e. about one-half of the corresponding value for a continuous weld, and in fact, similar to that for a fillet welded edge attachment (see below). If the ends of the weld were poorly finished this could be even lower.

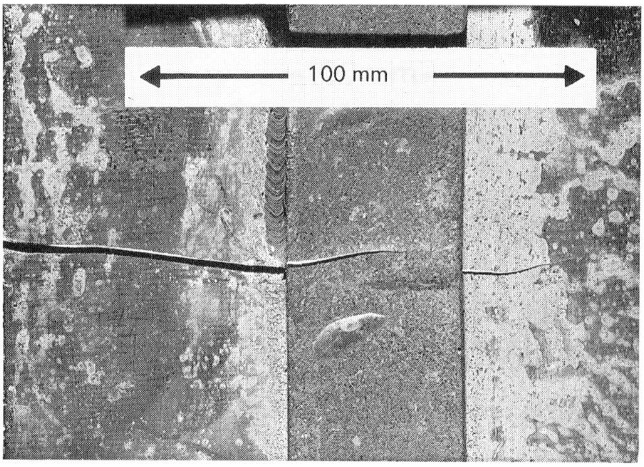

47 Fatigue cracking from end of tack weld attaching backing strip in longitudinal butt welded joint.

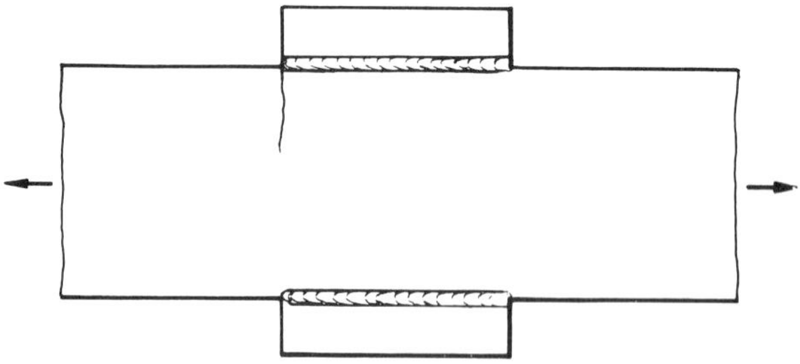

48 Fatigue cracking at attachment butt welded to edge of stressed plate.

The presence of the side plates produces a severe stress concentration which, if these specimens were machined from solid, would be relieved by a generous radius. With the welded joint the severe geometrical notch is further aggravated by the presence of the weld ends. Unfortunately welding offers a convenient means of making connections of this type, and it is imperative that the designer should become aware of these pitfalls.

Fillet welded connections

If the shape of the excess weld metal in transverse butt welds is sufficient to cause a reduction in fatigue strength, it can be expected that fillet welds, with their inherently more severe changes in shape, will be more sensitive to conditions of fatigue loading. Not only is this true but, because of the flexibility they allow in the arrangement of material in a structure, fillet welds easily permit the unthinking use of gussets, brackets, cover plates, and other odd attachments which may give rise to a general stress concentration in addition to the local effect of the weld.

Rather than considering transverse and longitudinal fillet welds separately, it is convenient to divide fillet welds into two other categories: non-load-carrying and load-carrying. Non-load-carrying fillet welds may be defined as attachment welds not designed to transmit the loads in the main members. Conversely, a load-carrying fillet weld transmits load from one member to another. This can be illustrated by considering a simple cruciform joint, Fig. 49. In Fig. 49a the fillet welds are not required to transmit any significant part of the load in the continuous member, where as in Fig. 49b all the load is transmitted through the welds and across the continuous member. This definition also applies

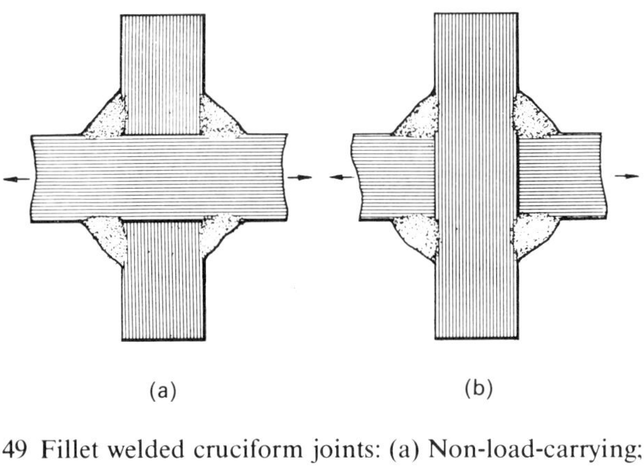

49 Fillet welded cruciform joints: (a) Non-load-carrying; (b) Load-carrying.

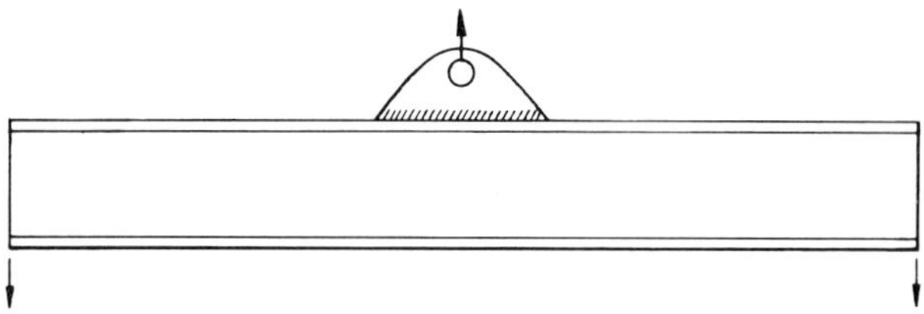

50 Lifting beam with fillet welded lug.

when the welds are parallel to the direction of stress. Some load will always be transmitted through a non-load-carrying joint, the amount depending on the size and geometry of the attachment. As will be seen later, the fatigue strength of non-load-carrying welds can vary significantly depending on the share of the load taken by the attachment.

There are many occasions in practice when fillet welds fall into both categories. For example, on a simple lifting beam the welds attaching the lug would certainly be load-carrying in transmitting the load applied to the beam, Fig. 50. They would, however, be non-load-carrying with respect to the bending stress induced in the beam flange.

Non-load-carrying fillet welds

Non-load-carrying fillet welds appear in a variety of forms, but in most cases they are associated with attachments of one sort or another, as illustrated in Fig. 51. As noted earlier, because a weld forms an integral part of the plate to which it is attached, fillet welds can introduce stress concentrations even though they are non-load-carrying. In this respect, the size effects discussed in Chapter 3 are particularly relevant and they go a long way to explaining the wide range of possible fatigue strengths. The extent of load transfer to the attachment increases with increase in its size and this causes an increase in the stress concentration at the weld toe or weld end. The important 'sizes' depend on the type of detail but of particular significance is the 'attachment length' L (see Fig. 34 and 51). Otherwise, the thickness of longitudinal attachments with their edges in contact with the stressed plate, Fig. 51c, or the width in the case of attachments with their surfaces in contact with the stress plate, Fig. 51d and e, are relevant. Main plate thickness is also important and, as in the case of increase in attachment size, an increase in main plate thickness can lead to a reduction in fatigue strength.

In a plate with a double fillet welded transverse attachment, the highest stress concentration occurs at the weld toes, and cracking will normally initiate at one of them and propagate through the plate, Fig. 51a and b. In terms of the nominal stress range in the plate, the fatigue strength at 2×10^6 cycles is about 95 N/mm^2, although test data are widely scattered. The actual fatigue strength depends upon the shape of the fillet weld, the extent of any undercut in the main plate at the weld toes and, of course, the plate thickness. An example of cracking

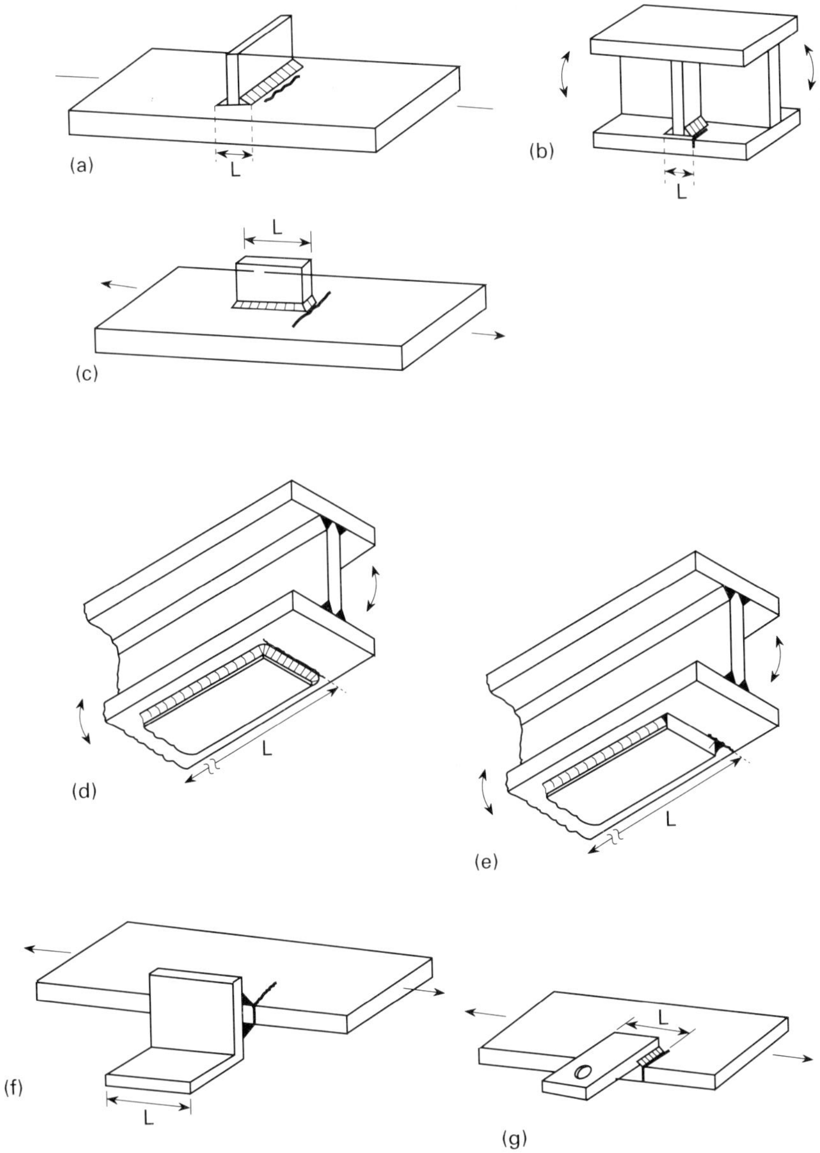

51 Examples of non-load-carrying fillet welded attachments.

from the toe of a fillet weld provides an illustration of the difficulty of detecting fatigue cracks, at least until they are very large, by which time most of the fatigue life of the joint concerned may have been consumed. Figure 52a shows a steel plate with a transverse fillet welded attachment, which has been subjected to fatigue loading. Visual inspection provides no indication that fatigue has occurred and it is only after checking with dye penetrant, Fig. 52b, that this is confirmed. The actual crack depth was 5.5mm, Fig. 52c.

In some instances, attachments are made using single fillet welds only. In such joints a high stress concentration also arises in the main plate of the weld root and fatigue cracking may initiate at either the weld toe or the root, Fig. 53. A problem with the latter is that fatigue

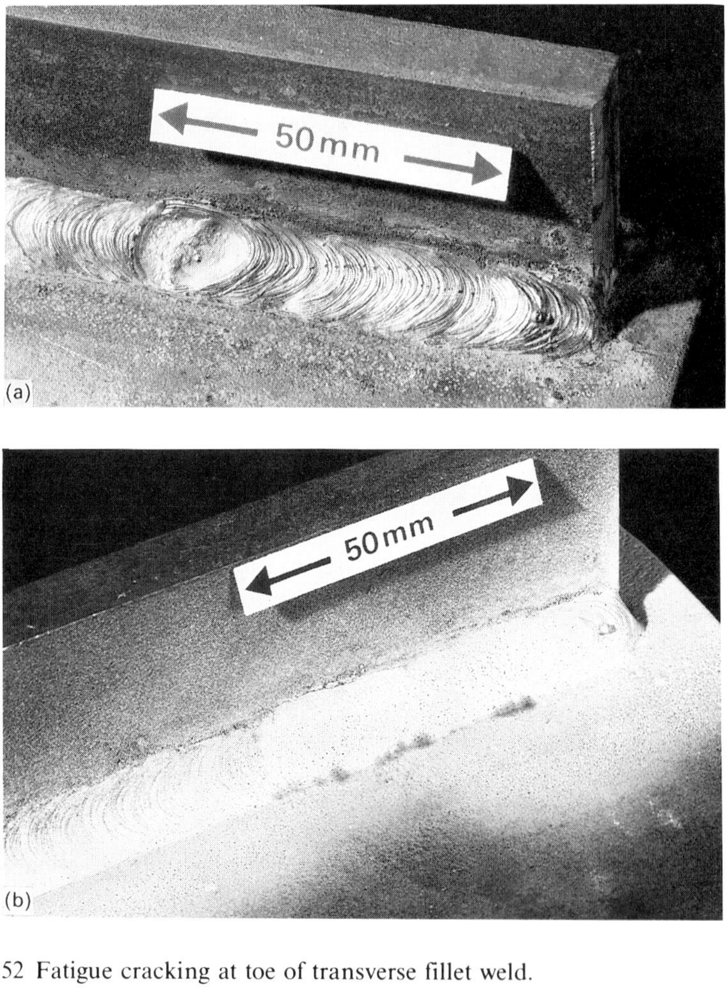

52 Fatigue cracking at toe of transverse fillet weld.

cracking is unlikely to be detected before final failure. Also, any measures taken to improve the fatigue strength of the joint by treating the weld toe (see Pt 2 for description of weld toe improvement techniques) will be ineffective.

There is no advantage to be gained in making these attachments with the welds parallel to the direction of stress, as in the cases shown in Fig. 51c and e. Cracking will initiate at the ends of the welds, Fig. 54, and, for short attachments, about the same fatigue strength will be obtained as for the transverse weld because, as noted earlier, the weld ends will produce a similar stress concentration to the weld toe. It is a fairly common practice to return the ends of the weld around the edge of the attachment, thus making short transverse welds. This again shows no improvement in fatigue performance, but does have the advantage of completely sealing the edges of the detail against possible corrosion. The fatigue strength of longitudinal attachments decreases with increase in attachment length, as illustrated in Fig. 35b.

A further reduction in fatigue strength occurs if the attachment is

(c)

both long and wide as in the case of a cover plate on a beam, Fig. 51d and e. For an extremely large attachment the average fatigue strength obtained at 2×10^6 cycles is only around 65 N/mm^2 for both transverse, Fig. 51d, and longitudinal, Fig. 51e, welds.

Attachments to the edges of stressed plates, Fig. 51f and g, have low fatigue strengths, comparable with those for cover plates, even if the attachment is relatively short. Based on the nominal stress range in the main plate, the fatigue strength at 2×10^6 cycles is about 70 N/mm^2, which is the same value as that already quoted for a plate with butt welded side plates, which has a similar geometry in the region of crack initiation, Fig. 55a. In contrast to longitudinal surface attachments which, in terms of fatigue strength, are unaffected by the addition of a transverse fillet weld around their ends, the continuation of the weld

53 Fatigue failure in plate with transverse attachment welded from one side.

around the ends of the edge type of attachment and across the edge of the main plate, Fig. 55b, reduces the fatigue strength of the detail to approximately 52 N/mm^2 at 2×10^6 cycles. Clearly, unless it is absolutely necessary to 'seal' the joint in this way, the practice should be avoided.

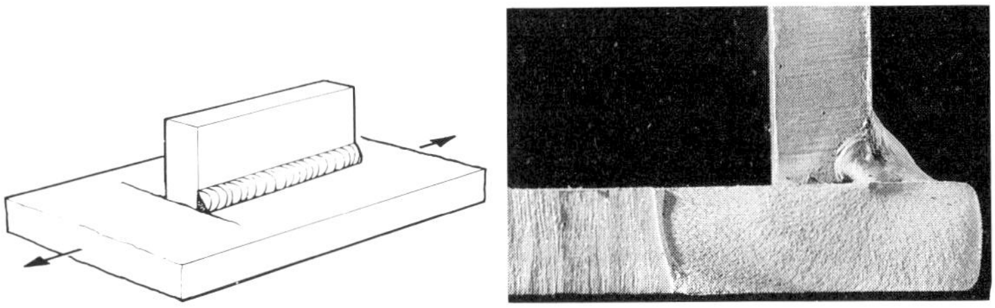

54 Fatigue failure from end of longitudinal fillet weld.

55 Fatigue cracking from end of fillet welded attachment to edge of stressed plate.

The low fatigue strengths of edge attachments compared with surface attachments can also be explained in terms of the size effect. In particular, the plate width, which is typically 100–150 mm in most specimens tested, is analogous to the thickness of the plate with surface attachments, typically 12 to 25 mm in most specimens tested. Thus, the condition produced is comparable with that in a very thick plate with a surface attachment and a correspondingly low fatigue strength is to be expected. Available test results, discussed in more detail in Pt 2, confirm this.

A final category of longitudinal non-load-carrying fillet weld includes intermittent fillet welds, such as those joining a web to a flange in a beam, or attaching a backing strip prior to making a butt weld, referred to earlier under 'longitudinal butt welds', Fig. 56. Although the local weld detail is the same as that in a longitudinal fillet welded attachment, the geometric stress concentration is lower, with the result that fatigue strength is slightly higher, around 120 N/mm^2 at 2×10^6 cycles. However, if the web to flange weld coincides with a cope hole, Fig. 57, the stress concentration is increased and the fatigue strength reverts to that of the longitudinal attachment.

Load-carrying fillet welds

In the types of welded detail discussed so far it has not been necessary to differentiate between the stress in the weld and the stress in the parent plate. In butt welds the nominal stress in the weld is the same as that in the members to be joined and in non-load-carrying welds the important stress is that in the main member to which the weld is attached.

56 Fatigue cracking from the end of a longitudinal intermittent web/flange fillet weld.

For details containing load-carrying fillet welds, however, the nominal stress in the weld and that in the members being joined can be different. The reason for this may be explained by examining how fillet weld sizes are normally calculated.

With reference to Fig. 58, it is assumed that the load, P, is carried on the total weld throat area, $2A_w$, where A_w is equal to the length of the weld times the throat thickness, T. The weld is dimensioned so that $P/2A_w$ = the allowable design stress for the weld and, knowing the length

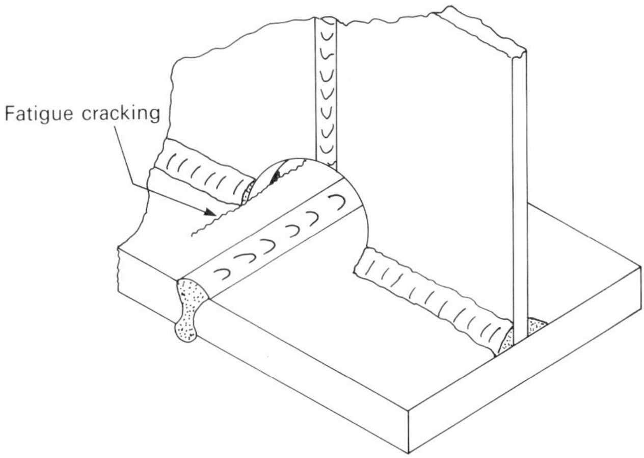

57 Fatigue cracking from the end of a web/flange fillet weld at a cope-hole.

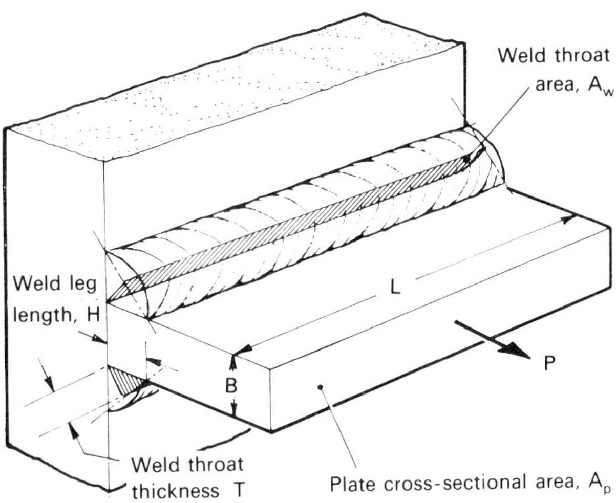

58 Dimensions in transverse load-carrying fillet welded joint.

of the weld, the throat thickness, T, and hence the weld leg length, H, can be derived. For a normal 45° fillet, H is taken as $\sqrt{2}.T$.

The ratio plate/weld stress, i.e. $2A_w/A_p$, is termed the design ratio. For example, under static loading conditions the allowable weld stress may be 115 N/mm^2 and the allowable tensile stress in the plate 155 N/mm^2, which would give a design ratio of 1.35 if the allowable stresses in both components were satisfied. The designer may choose to develop the same stress in both the plate and the weld, that is a design ratio of one. In this situation each fillet weld leg length would be equal to $\sqrt{2}$ times half the plate thickness (H=0.7B).

The same calculation procedure is used for other types of fillet welded joint. For example, in a double lap joint with longitudinal fillets, Fig. 59, the weld throat area is simply the product of the total length of weld 4 times the weld throat thickness. If the welds were continued across the ends of the cover plates this would constitute an additional length of weld and would reduce the nominal weld stress, but would have no effect on the nominal stress in the plates.

The design of load-carrying fillet welded joints for static loads is reasonably straightforward but, unfortunately, their behaviour under fatigue loading conditions is rather more complicated. Many different weld configurations and joint forms may be used, all of which might show slightly different fatigue strengths and various modes of failure. The principal factors involved can be shown by considering one or two of the more usual types of joint.

Examination of the stress flow pattern in a cruciform joint reveals that there are two basic stress concentrations from which fatigue cracks may initiate, Fig. 60. As the stress has to flow through the weld itself, severe internal notches are formed by the ends of the plates at the weld roots, in addition to the now familiar notch effect at the weld toes. The question thus arises as to which point of stress concentration will be the most severe and hence determine the behaviour and strength of a fillet welded joint of this kind.

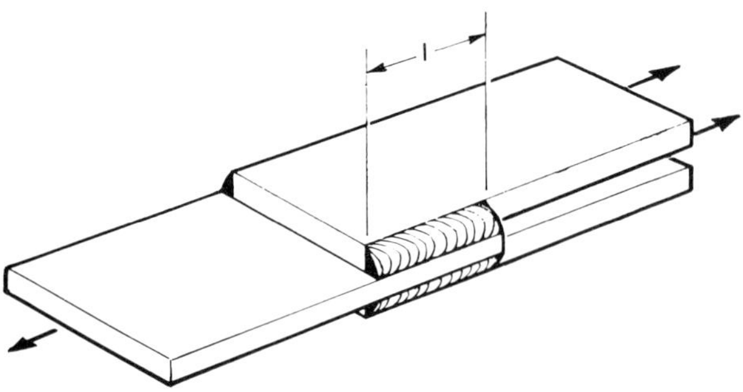

59 Longitudinal load-carrying fillet weld.

Taking the cross-section shown in Fig. 61(a), it can be imagined that the stress concentration produced at the weld toes on the stressed plates is not much affected by decreasing the fillet leg length from H_1 to H_2, in that the actual change of section at the toe remains the same. On the other hand, the stress on the weld throat will be increased significantly, providing an opportunity for the root notch to initiate a crack as shown. In contrast, cracks might preferentially be located at the toes of the larger fillets. It follows that the location of failure depends on the ratio leg length/plate thickness, (H/B).

Optimum performance is achieved when there is just enough weld metal (H is just sufficiently large) to ensure that failure would be from the toe rather than the root of the fillet weld. Design stresses for the plate and weld throat are based on this requirement and, when applied to the cruciform joint, they give the general result that leg length is approximately equal to plate thickness (H = B). For example, typical permissible stresses for a life of 2×10^6 cycles would be 85 and 57 N/mm^2 for the plate and weld respectively, so that

$$\frac{2A_w}{A_p} = \frac{85}{57} = 1.5 \text{ (design ratio)}$$

and, from the calculations given previously, if B = 15 mm, T = 0.75B, from which H = 16 mm.

A thicker plate would require a correspondingly larger weld size.

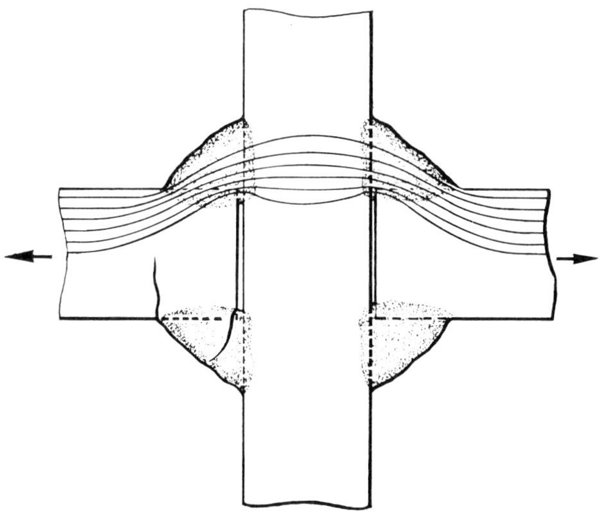

60 Two possible sites for fatigue cracking in transverse load-carrying fillet welded joint.

Bearing in mind, however, that weld metal volume increases at H^2, an alternative approach might be the use of 'partial penetration' welds, Fig. 61b. As indicated, it is possible in such welds, by virtue of the positive root penetration, p, to achieve a given throat dimension with a leg length smaller than would be required in the normal fillet weld, and this may be taken into account. The design of joints with partial penetration welds is discussed more fully in Pt 2, but, as an illustration, the leg length of 16 mm quoted above (for B = 15 mm) could be reduced to 11 mm if 3 mm depth of penetration, p, could be assured (and it is to be noted that 'assurance' is important).

Increasing the weld area above that which represents optimum design, by the use of larger or partial penetration welds, will not give a further improvement in the fatigue performance of the joint unless the penetration is increased to such an extent that a full penetration butt weld is formed, Fig. 62. This eliminates the stress concentration at the weld root and tends to reduce concentration at the toes. Even so, the basic cruciform configuration is retained and the improvement of

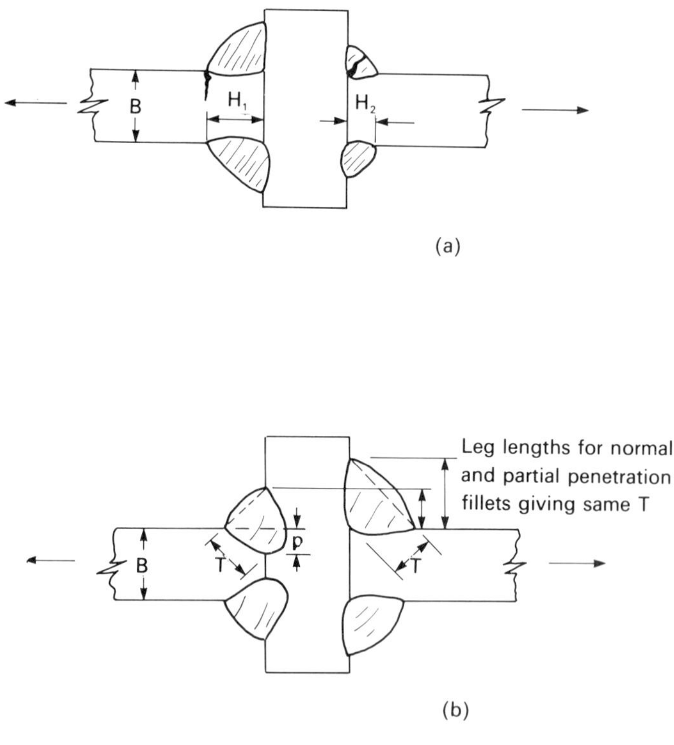

61 Load-carrying transverse fillet welded joints.

strength somewhat limited; for a fatigue life of 2×10^6 cycles the permissible stress range for the plate would be 95 N/mm^2 compared with 85 N/mm^2 for the fillet welded detail. This is not to say, however, that full or partial penetration welding offers little advantage; for thick plate assembly the question of fabrication cost represents an additional consideration.

A fillet welded T joint, Fig. 63, is, in effect, half a cruciform joint and would exhibit similar behaviour and fatigue strength from the point of view of failure in the weld or in the stem plate. But, as shown, a third mode of failure is possible if bending stresses are induced in the transverse member, i.e. cracking from the weld toes on that member. It follows that the magnitude of such bending stresses must be accounted for and, indeed, must be effectively reduced before any overall improvement of joint strength can be secured by the introduction of a full penetration butt weld.

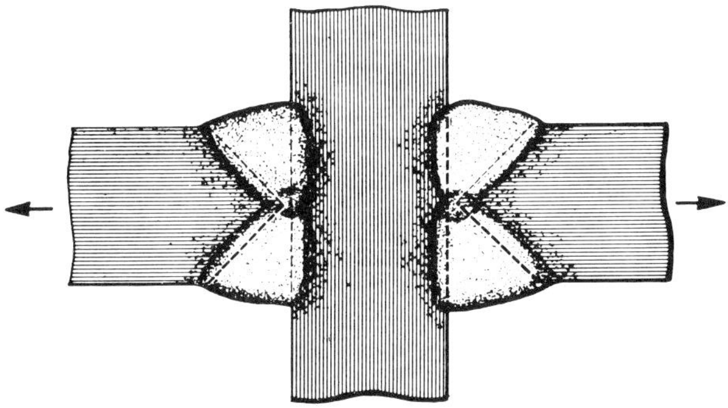

62 Transverse butt welded cruciform joint.

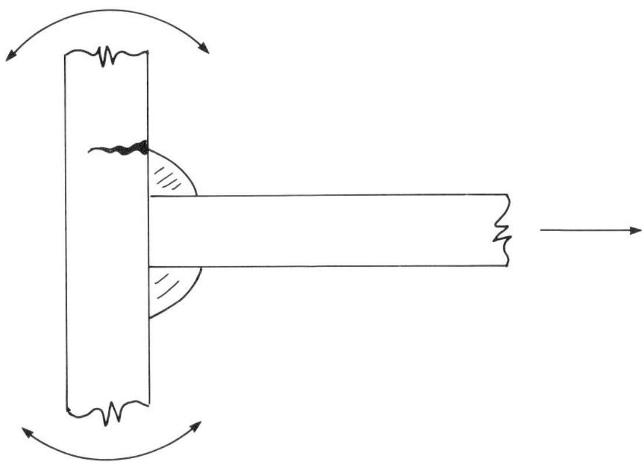

63 Load-carrying fillet welded T joint.

One feature of cruciform joints which will not be present in the T-joint is misalignment of the loaded plates. As was the case with transverse butt welds, misalignment in cruciform joints subjected to axial loading leads to the introduction of secondary bending which increases the stress range at the weld toe and, in the case of fillet welds, at the weld root. As a result, the fatigue strength of the joint is reduced, as illustrated for the two modes of failure in Fig. 64. As before, correlation of the results is achieved by expressing them in terms of the total

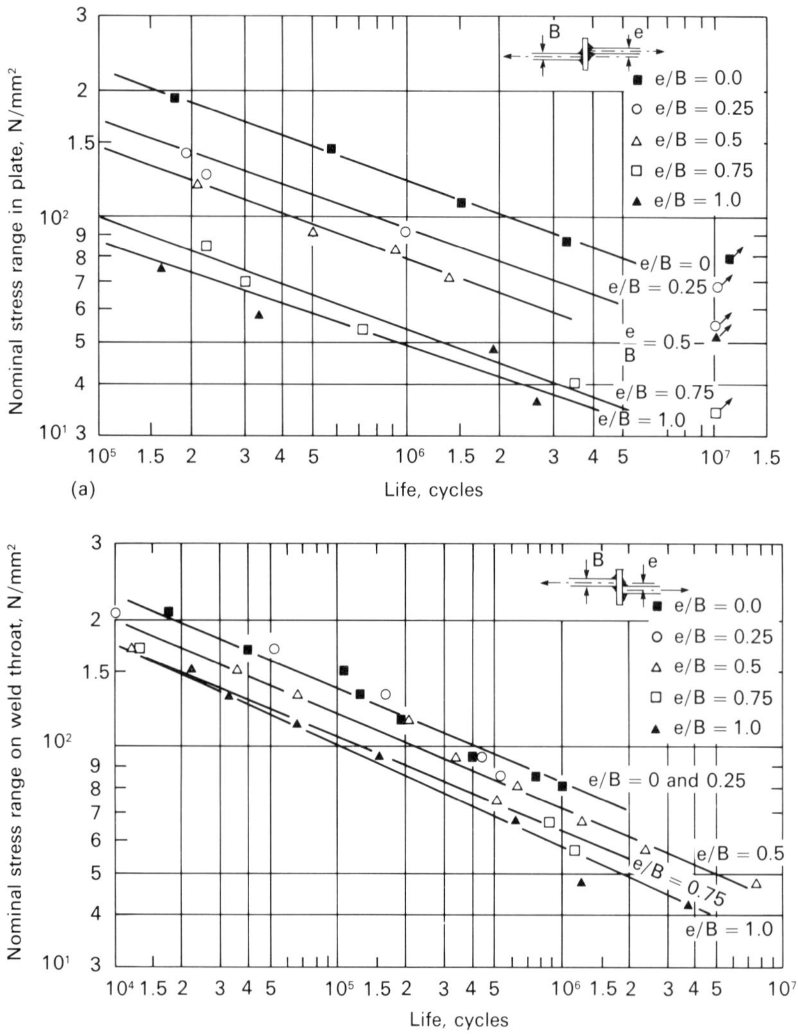

64 Fatigue test results for misaligned cruciform joints: (a) Joints failing from the weld toe in the plate; (b) Joints failing from the weld root in the weld throat.

stress range. It will be noted that misalignment has less effect in the case of the fillet welds failing in the weld throat because the site of fatigue crack initiation, the weld root, is closer to the neutral axis of the section with respect to the induced bending moment. It is to be expected that some of the scatter observed in published fatigue test data for cruciform joints is due to misalignment but, since investigators rarely consider misalignment, the effect cannot be quantified. With regard to joints in real structures, there is even more likelihood than was the case with transverse butt welds that bending due to misalignment will be inhibited as a result of restraint, particularly from the cross plate which, in the fatigue test specimen, offers no restraint.

Longitudinal load-carrying fillet welds, such as those which might be used to make a double lap joint, Fig. 65a, are less sensitive to weld size since the changes of section at the weld toe and weld root become parallel with the direction of applied stress and do not, therefore, act as

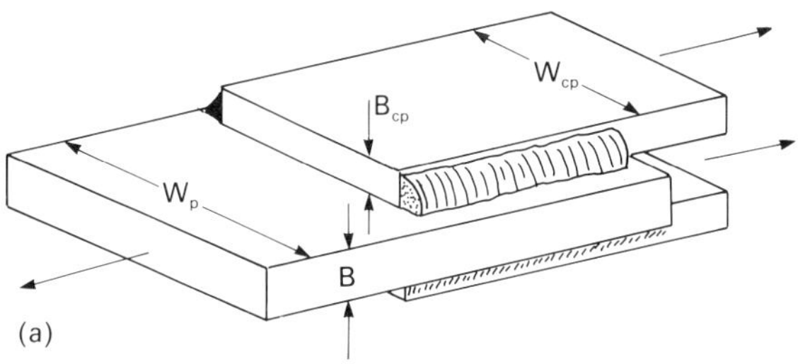

(a)

(b)

65 Lap joint made using longitudinal fillet welds: (a) Dimensions; (b) Fatigue failure from weld ends in cover plate.

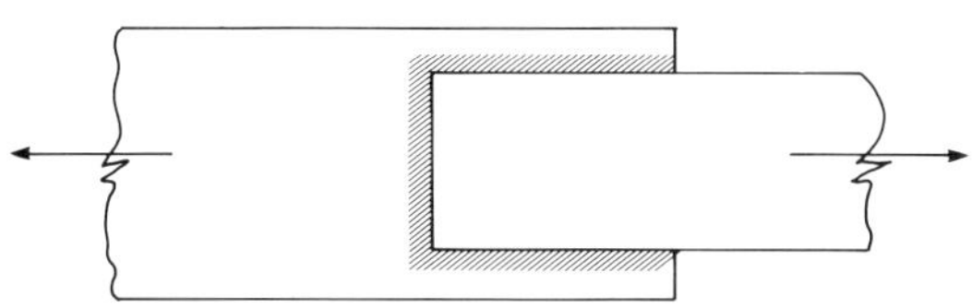

66 Lap joint with weld continued around end of cover plate.

points of stress concentration. The weld ends do, however, create a local condition of stress concentration in both the main and cover plates and provide sites for crack initiation. In this respect it is necessary to note that, in general, welds on the edges of a stressed member can be more damaging than those located on a surface. If, therefore, the double lap joint is so designed that the same nominal stress is produced in the main and cover plates ($B \times W_p = 2B_{cp} \times W_{cp}$), failure will initiate from a weld end and propagate into the cover plate, Fig. 65b, giving a fatigue strength of about 66 N/mm^2 for 2×10^6 cycles. Some improvement over this value can be obtained by welding the ends of the cover plates as well as each side, Fig. 66, and proportioning the joint so that failure would occur preferentially at the transverse weld toe, i.e. $2B_{cp} \times W_{cp} > B \times W_p$. In this form, the fatigue strength at 2×10^6 cycles would be about 85 N/mm^2, expressed in terms of the stress range in the main plate.

For this kind of joint it is necessary to determine the minimum weld throat area required, and it will be noted that for the same overlap dimension the combination of longitudinal and transverse fillets gives a greater length of weld, hence a possible reduction of weld size. It may also be necessary to account for the problem that the width of the main plate is much larger than is reasonably required for cover plates. In this case, use is made of an arbitrary design rule which, for stress calculations, states that an 'effective' width of main plate shall be taken, Fig. 67.

In view of the poor fatigue properties of non-load-carrying welds on plate edges, it is not surprising to find that a similar detail in a load-carrying joint has an even lower fatigue strength than lap joints, being about 50 N/mm^2 at 2×10^6 cycles. Again, as in the non-load-carrying joint, the deposition of a sealing weld at the ends of the edge plates further reduces the strength, to around 44 N/mm^2 at 2×10^6 cycles. There do not appear to be any circumstances where details of these types cannot be avoided and it is strongly recommended that they are not used.

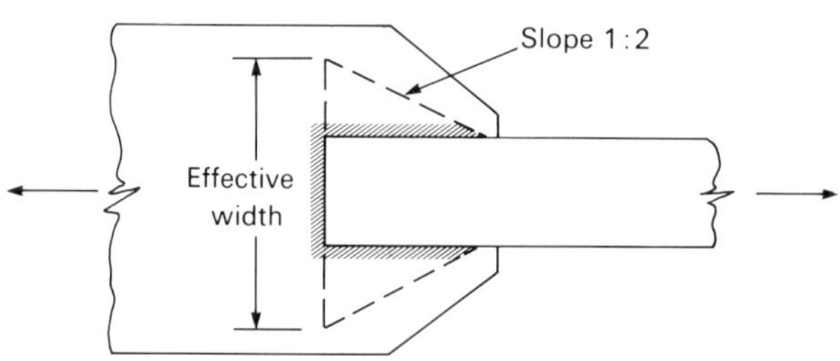

67 Proportion of plate width assumed to be effective in carrying load transmitted from cover plate in lap joint.

It will have been observed that the fillet welded joints discussed have all had symmetrically disposed welds and components. Although the associated fatigue strengths are quite modest, they are dependent on such a configuration. In contrast, a joint containing a single load-carrying fillet weld is likely to exhibit very poor strength since it will be subjected to severe local stress; similarly, single lap joints in flat plate construction will be exposed to secondary bending stresses which will also reduce fatigue performance, unless the joint is supported by additional stiffening material. The selection of joint form for fatigue service is thus a crucial aspect of welding design.

The fillet welds so far considered have all tended to have poor fatigue strengths, but this is not always so. It is the stress concentrations in the loading path which are important and, in this respect, transverse and discontinuous longitudinal fillet welds introduce severe concentrations, at the weld toe and weld end respectively. On the other hand, the continuous longitudinal fillet weld, as in the web-to-flange joint of a beam, Fig. 68a, presents no change of shape in the way of the main (bending) stress field and gives good fatigue strength. In welds of this kind made by the MMA process, fatigue cracking usually initiates on the weld surface at stop/start positions, Fig. 68b, which represent minor discontinuities. With automatic weld-

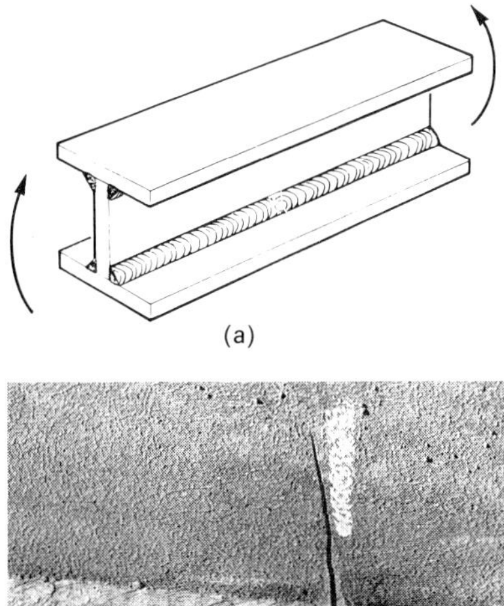

(a)

(b)

68 (a) Continuous weld-to-flange fillet weld; (b) Fatigue failure initiated at stop/start position.

ing even these blemishes are absent (provided the weld is deposited continuously) and cracking is initiated by surface ripples, or possibly embedded flaws. The fatigue strengths, expressed in terms of stress range on the plane of the weld, e.g. the bending stress on that plane, for a life of 2×10^6 cycles are about 140 and 160 N/mm^2 respectively. It can often be demonstrated that intermittent fillet welds would be sufficient to carry the shear load in a web-to-flange joint, but these at once reintroduce the weld end notch effects, capable of initiating cracks in both the web and flange of a beam. In comparison with the values quoted above, the fatigue strength at 2×10^6 cycles would be reduced by 20–45%.

Welded structures

So far, consideration has been given to the stress concentrations associated with the welded joint itself, and fatigue strengths have been expressed in terms of the nominal stress in the region where fatigue cracking will be initiated. There is, however, a further consideration when joints are incorporated in an engineering structure: it might be necessary to have to account for their exposure to stress concentration arising from the configuration of the assembly. A simple example illustrates this point.

Nominal stress range in the welded, flat plate detail, Fig. 69a, is readily determined as $S = P/BW$, and a permissible value, say S_D, is obtained

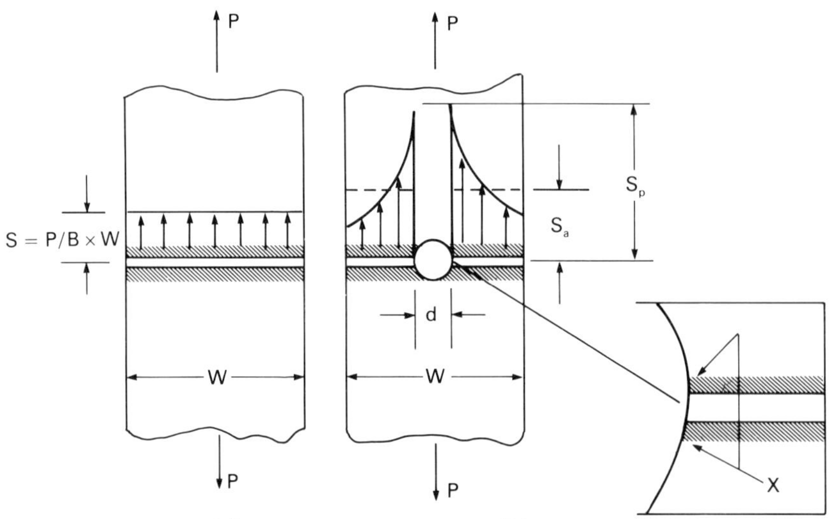

69 Fillet weld in region of stress concentration stemming from hole.

by reference to design data for the transverse, non-load-carrying fillet weld (S_D being dependent upon required life, N). Suppose, however, that for a functional purpose the plate contains a round hole with attachments placed on each side, as shown in Fig. 69b. Average stress on the net section is $S_A=P/B(W-d)$, which might conventionally be taken as uniformly distributed, but for fatigue loading conditions it is, of course, essential to recognise the non-uniform distribution caused by the presence of the hole.

As previously explained (see 'Stress concentrations'), the important feature of actual distribution of stress on the net section is the existence of peak stresses, S_p, in material immediately adjacent to the hole since these, in effect, represent the magnitude of stress range experienced by the affected material with each load application. Moreover it is necessary to make the assumption that, in the welded detail, the weld toes in the same regions (marked 'X') will also be acted upon by S_p. Thus it follows that, if the same life, N, is to be obtained, S_p should not exceed the appropriate value of S_D. In consequence, it would be necessary to increase the net section area by a considerable amount.

In principle the cross-sectional area to be provided can be deduced from the requirement that $K_t.S_A \leqslant S_D$, given that a value for K_t (S_p/S_A) can be inserted. Such values are established for a range of geometrical forms which introduce the condition of stress concentration and it has been mentioned that, for round holes in a tensile stress field, K_t may approach 3.0. Clearly, a very large increase of net area would be necessary for the perforated plate detail, arising directly from the coincidence of a weld toe notch effect and a stress concentration produced by general configuration.

It follows that, if fatigue loading is a design criterion, such a condition should be avoided whenever possible. In practice there will be many apparent difficulties in the application of this principle, and where these are quite real it may not be easy to adopt the quantitative analysis because of uncertainty about the correct value of K_t. Examples of this problem are found in circumstances where the disposition of members is such that one suffers a marked change of stiffness, as in the shell plating of a thin-walled tank at that point where it is suddenly supported by the end of stiff cradles, Fig. 70. It can be visualised that the load in a full tank would lead to high stress concentration in the shell plate and would take the form of a fluctuating inertia load if the tank were mounted on a road vehicle. Laboratory or prototype testing is often carried out to arrive at a solution to problems of this kind (and it will be noticed that road tankers invariably have a girth strap inserted between the cradles and shell plating to make change of stiffness more gradual).

In other fields of engineering it is essential to support design work with experimental data for specific forms of structure. Joints in tubular connections or nozzles in pressure vessels can, for example, exhibit a wide range of K_t values, depending on the configuration and relative size of the members, and these need to be explored experimentally to be able to make relevant stress calculations. Nevertheless, there are opportunities in the broader context of general design to take up the principle of avoiding, or ameliorating, the effect of a combination of stress concentrations.

One particular case could be that of a framework in which beam sections meet at right angles, creating stress concentration at the 90° junction of flanges, in which zone there is also an additional notch in the form of the end of a butt weld, Fig. 71a. The detail produced is similar to the butt weld on a plate edge which has already been discussed. The addition of gussets of a form permitting the weld ends to be blended into flange edges by postweld dressing, Fig. 71b, would ease the severity of each of the stress concentration features and encourage reduced disturbance of stress flow.

The same end result could, in fact, be sought for the perforated plate, Fig. 69b, if design requirements made it necessary to have both a hole and the coincident attachment. Provided the hole did not have to be round, an elliptical one with its long axis parallel to the direction of applied stress could be designed to give at least a noticeably reduced value of K_t, once more by causing a smaller disturbance of the stresses in the region of the discontinuity; even so, a strong question mark should remain against the necessity of placing an attachment in such a vulnerable position! Further aspects of design application are pursued in Pt 2.

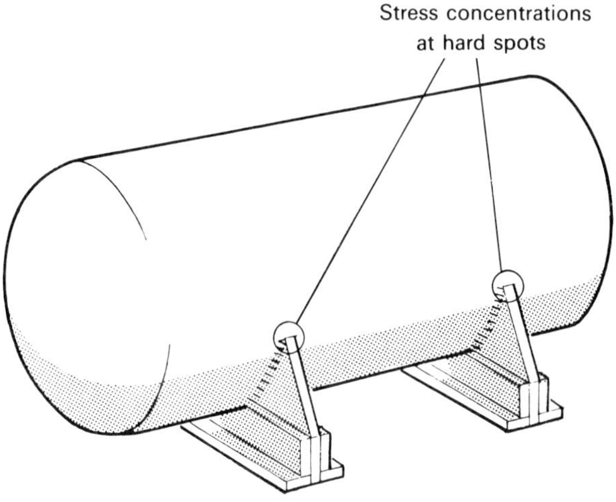

70 Fillet weld in region of stress concentration arising from change of stiffness.

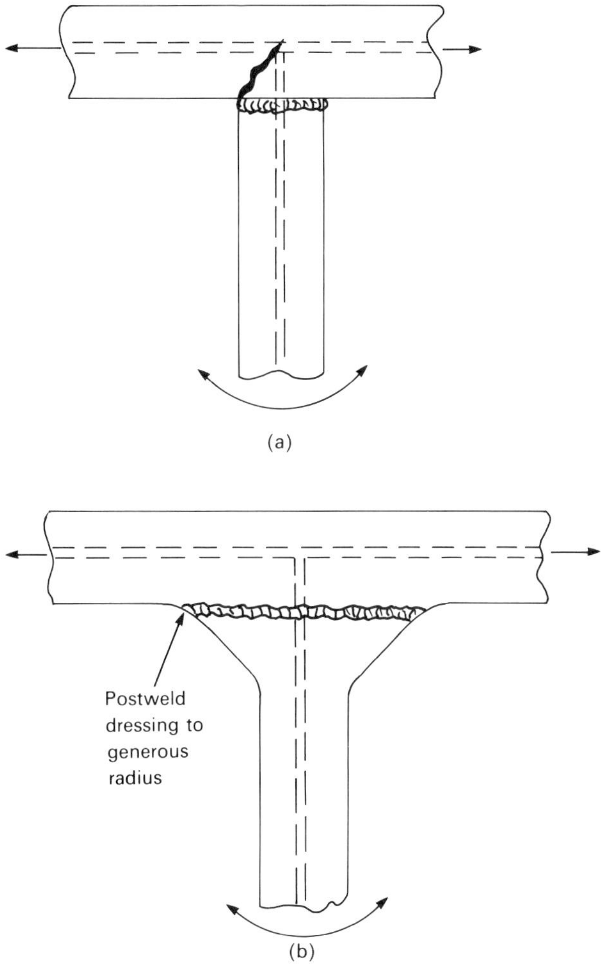

71 Improvement in design of cross-girder connection.

Chapter 5

Conclusions

The discussion of the general fatigue behaviour of welded joints has shown that their fatigue strengths are dominated by geometric factors. In most cases, the fatigue lives are spent propagating cracks from pre-existing crack-like flaws, the process of fatigue crack initiation occupying an insignificant part of the fatigue life. This has a profound effect on their fatigue behaviour since factors which affect fatigue crack initiation may have quite different effects on fatigue crack propagation.

Fatigue strengths are considerably lower than the fatigue strengths of unwelded components, and factors which might prove beneficial for unwelded components, because of the greater significance of crack initiation, have no effect on the fatigue strengths of welds. A notable example is the effect of material strength, and the choice of a higher strength material to solve a particular fatigue problem will not normally be successful in the case of welded joints. Overall, welded joint geometry is also significant and it accounts for the wide range of fatigue strengths obtained from different weld details, even those in which the life consists of the propagation of pre-existing flaws. This variation in fatigue strengths is summarised in Table 1, which lists some typical welded joints in steel for which sufficient test data have been obtained to enable S–N curves to be derived, together with the respective average stress ranges corresponding to fatigue lives of 10^5 and 2×10^6 cycles.

The presence of high tensile residual stresses in welded joints is also significant in that applied compressive stresses are just as damaging as applied tensile stresses. In most practical circumstances, this problem cannot be alleviated by thermal stress relief.

Weld quality may also be important but, in general, welding imperfections are only significant if they introduce stress concentrations which are more severe than those already present as a result of the basic weld geometry.

Thus, the avoidance of fatigue failure is very much the province of the design engineer, including the wise choice of weld details to optimise fatigue strength, recognition of potential fatigue problems

Table 1 Summary of typical fatigue strengths for steel

Detail	Sketch of detail and mode of fatigue cracking	Fatigue strength (R = 0) N/mm^2	
		10^5 cycles	2×10^6 cycles
Plain as-rolled C–Mn structural steel plate	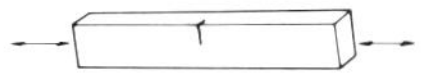	300	200
Continuous automatic longitudinal butt and fillet welds (no stop/starts)	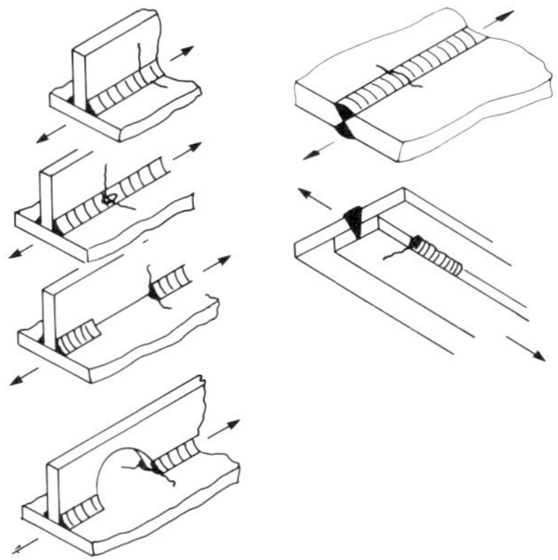	350	160
Continuous manual longitudinal butt and fillet welds		325	140
Longitudinal butt weld on tack welded backing strip		280	125
Intermittent longitudinal fillet weld		260	120
Web/flange weld at cope hole		230	95

Table 1 cont'd

Detail	Sketch of detail and mode of fatigue cracking	Fatigue strength (R = 0) N/mm²	
		10^5 cycles	2×10^6 cycles
Fillet or butt welded attachments to the edges of stressed members	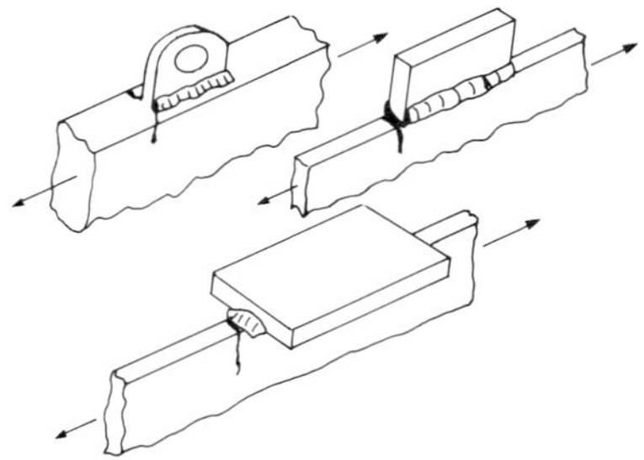	180	70
Full penetration T or cruciform joints	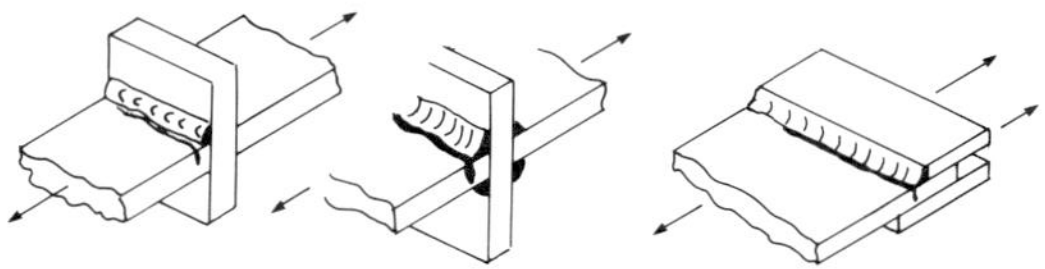	360	95
Fillet welded T or cruciform joints		250	85
Lap joint with transverse fillets		250	85
Weld throat failure in transverse weld (based on stress on weld throat)	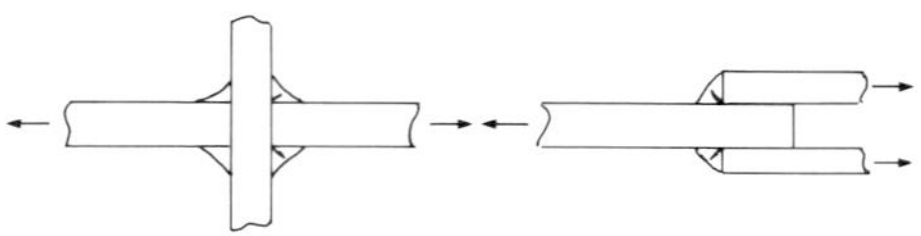	154	57

Table 1 cont'd

Detail	Sketch of detail and mode of fatigue cracking	Fatigue strength (R = 0) N/mm²	
		10^5 cycles	2×10^6 cycles
Transverse butt welds with good profile	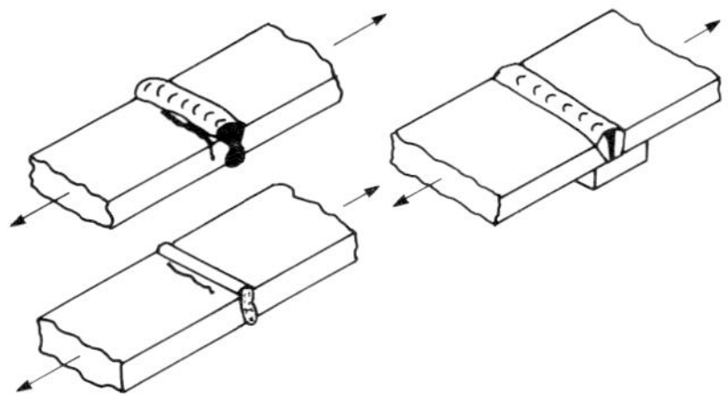	355	155
Transverse butt weld with poorer profile (including submerged-arc)		260	116
Transverse electron beam butt weld		310	155
Transverse butt weld on a permanent backing strip		260	100
Fillet or butt welded stiffeners or attachments to the surfaces of stressed members	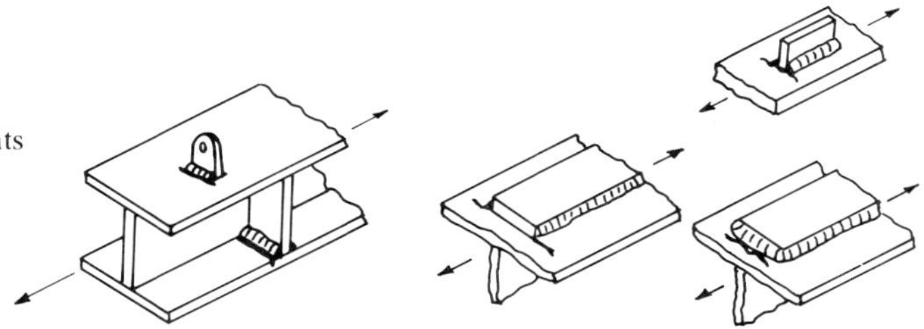	250	95
Cover plates on beam flanges		160	65

Table 1 cont'd

Detail	Sketch of detail and mode of fatigue cracking	Fatigue strength (R = 0) N/mm²	
		10^5 cycles	2×10^6 cycles
Lap joint with longitudinal fillet welds:	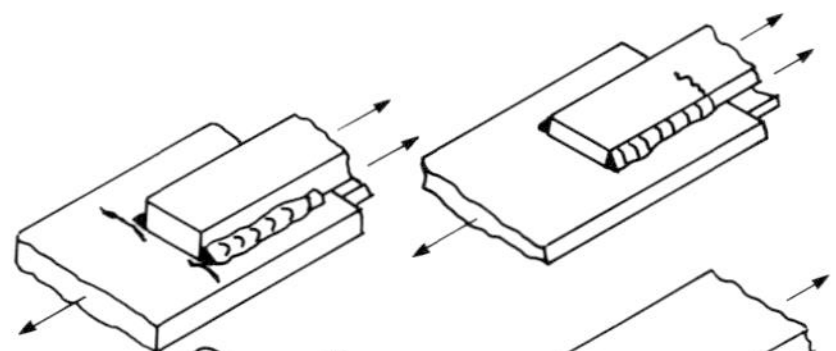		
(i) crack initiates on surface		230	80
(ii) crack initiates on edge of cover plate		200	66
Load-carrying weld on plate edge:	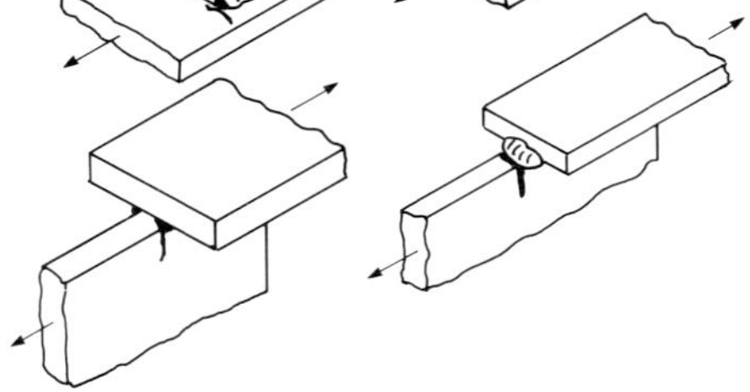		
(i) not welded around end		150	50
(ii) welded around end		125	44

associated with welded joints and full appreciation of the fatigue loading to be experienced by the structure. This emphasises the need for clear design guidance. Fortunately, comprehensive design rules are available to cover a wide range of welded joint types and these are described and their application discussed in Pt 2 of this book.

Part 2

2 Design rules

Chapter 1

Fatigue design rules for welded steel joints

Introduction

Fatigue failures are avoided by ensuring that all features of a structure (e.g. welded joints) which will experience fatigue loading in service have adequate fatigue strengths. In the case of welded structures, the most widely used method for performing fatigue checks at the design stage is to refer to S–N curves for the relevant weld details, and this is also the basis of fatigue design rules.

Fatigue design rules for welded structures have developed extensively over the last twenty years, following a major review and statistical analysis of published fatigue data and the publication of rules by The Welding Institute. Initial effort was directed at steel bridges and the rules in BS 5400 evolved from those published by the Institute. Later, as a result of the rapid exploitation of oil and gas fields in the North Sea, a number of major European research projects provided design data for tubular structures, which, together with the welded plate design data contained in BS 5400, were the basis of new fatigue design rules for offshore structures. Broadly the same design rules have since been adapted for application in pressure vessel design and for considering fatigue of aluminium alloy welded structures. At the same time, considerable international collaboration has taken place, leading to design rules which are in good agreement with many National codes. This has been possible because of the general acceptance of certain basic principles upon which fatigue design rules for welded joints should be based. These principles are discussed in this chapter, which also presents the design rules for steel welded joints and considers their practical application.

A factor which has played an enormous part in the development of the design rules, both in terms of their form and content, was the realisation that in most cases fatigue cracks in welded joints originate from inherent welding flaws and therefore that the fatigue life consists mainly of the propagation of a crack. Some of the implications of this were considered in Pt 1. Implications related to the design rules require some understanding of fatigue crack propagation behaviour, expressed in fracture mechanics terms. Appendix III is devoted to this subject and reference to that appendix before studying the rules should provide some insight into this background. In particular, it will be clear why most of the design S–N curves are parallel, why there is a thickness penalty and why there is a distinction between the design S–N curves for some joints based on their relative dimensions.

Basis of rules

The design rules are based on a statistical analysis of fatigue test data obtained from tests on welded joints, to produce the best-fit mean S–N curve by the method of least squares. A straight line relationship between log S and log N was assumed, where S = stress range and N = fatigue life. The standard deviation of log N, σ, which is a measure of the scatter in fatigue lives obtained, was also calculated. An S–N curve two standard deviations below the mean (M–2SD), which, assuming a log-normal distribution of fatigue life, is the lower 95.4% confidence limit and represents approximately 2.3% probability of failure or a 97.7% survival limit, is taken as a reasonable basis for design, but other curves (M, M–1SD) are defined, Fig.1.

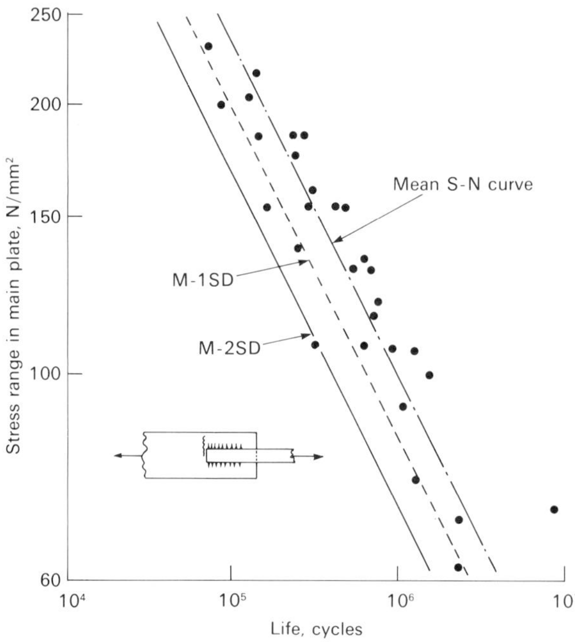

1 Typical set of fatigue data for welded joint in steel.

Design stresses for a given welded joint are represented by a single S–N curve (the M–2SD curve), taken to be applicable **irrespective** of the applied stress ratio. In other words, the values quoted refer to **permissible stress range**, which remains the same whatever the magnitude of the applied mean stress. The underlying reason for the dominant effect of stress range alone, connected with the influence of residual stresses, is explained in the general survey of fatigue behaviour of welded structures given in Pt 1. It should, however, be emphasised that this stress range basis of design does not deny the validity of general engineering fatigue rules which take account of mean stress; the concept is peculiar to the welded joint in the 'as-welded' condition.

Joint classification

For the purposes of fatigue design each part of a welded joint which could act as a site for fatigue cracking must be placed into one of nine classes, designated A, B, C, D, E, F, F_2, G, and W. The classification of joints is set out in Table 1 in terms of eight groups of details. Different details are combined in a single class on the basis that they exhibit similar fatigue strengths. Class A contains only unwelded plate. It is recognised that, except in respect of plate with flame cut edges, tensile residual stresses will not be present in such details and the stress range design approach used for welded details is over-conservative. However, in practice, unwelded areas of a welded fabrication are unlikely to limit the fatigue strength of the structure so that the choice of design stress is not critical. The class is included here merely for completeness, but design stresses are not given.

The fact that Class A does not include any welded details may appear to contradict the observation that fully machined butt welds can have a fatigue strength which is as high as the parent plate. However, for design purposes account is taken of the fact that, in practice, the fatigue strength of such welds can be reduced by the presence of welding defects which are too small for detection by currently available non-destructive testing methods.

In addition to a description of each joint, Table 1 contains some explanatory comments and indicates the relevant stress and its direction and likely potential sites for fatigue cracking. The use of the joint classifications and associated design stresses presented here is intended to prevent such cracking leading to failure. However it is the site for fatigue cracking which determines the classification, not the joint type. Thus it will be apparent that a particular joint could fall into more than one class, depending on the way in which it may fail by fatigue. In general, three possible sites for fatigue cracking should be considered: the toes of the weld in each of the members joined, giving

Table 1 Classification of details in welded components and structures

Group 1 Plain material free from welding

Potential fatigue failure modes

In plain material, fatigue cracks initiate at the surface, usually at either irregularities or corners. In welded constructions, fatigue failure will rarely occur in plain material, since the fatigue strengths of welded joints will usually be much lower. In members with rivet or bolt holes or other stress concentrations arising as the result of the shape of the member, fatigue cracking will usually initiate at the stress concentration.

Repair welds

If plain material is repaired by welding (e.g. surface irregularities in forgings), the region is reclassified as a transverse butt weld (see Group 4).

Description	Manufacturing requirements	Type No.	Class	Explanatory comments	Examples showing relevant mode of fatigue cracking, stress direction and location
Plate or structural section of constant or smoothly varying cross-section with no holes or re-entrant corners.	All surfaces and edges fully machined and polished smooth. No flame cut edges.	1.1	A	Beware of assuming Class A or B for a member which may acquire stress concentrations during its life, e.g. corrosion pitting, surface damage. In such an event, assume Class C or lower.	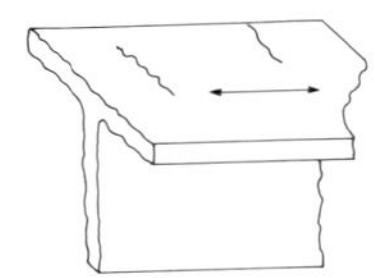
	Edges as rolled, extruded or machined smooth. Flame cut edges subsequently machined or ground smooth in the direction of stressing.	1.2	B		
Plate or structural section, including members with apertures or re-entrant corners.	Any flame cut edges subsequently machined or ground smooth in the direction of stressing.	1.3	B	Allowance must be made for stress concentration at apertures or re-entrant corners by designing on the basis of the net section stress range multiplied by the relevant SCF.	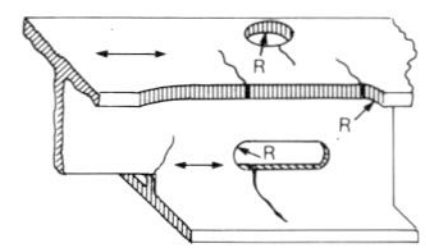
	Any cutting by planing or machine flame cutting with procedure controlled to ensure freedom from cracks.	1.4	C		

	Small holes drilled or reamered (may contain bolts for minor fixtures).	1.5	D	Based on net section stress range; the classification already allows for the SCF due to the hole.	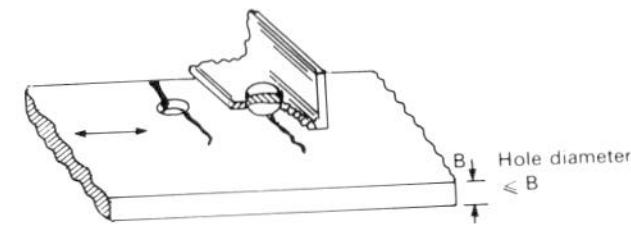
Crotch corner at branch connection (nozzle) in pressurised vessel.	Free from welding.	1.6	C	Stress must include relevant SCF.	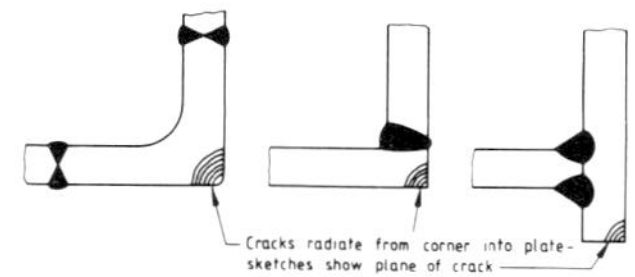
	Any repair welds dressed flush and proved free from significant flaws by appropriate NDT.	1.7	D		

Group 2 Butt welds between continuous plates essentially parallel to the direction of applied stress

Potential fatigue failure modes
With the weld overfill dressed flush, the fatigue performance will be critically dependent on weld quality, welding flaws being the most likely sites for fatigue crack initiation. Indeed, in most practical cases flaws which may reduce the potential fatigue life are too small to be detected reliably using standard NDT techniques. Thus, it is not usually possible to design dressed flush butt welds to their full potential. In as-deposited joints, fatigue cracks might also initiate at flaws, or on the weld surface at surface ripples, lumps due to stop/starts or as a result of an irregular root profile.

Significance of welding flaws
In all cases, it is necessary to prove that the joint is free from significant welding flaws by appropriate NDT. Fitness-for-purpose is a basis for assessing detected flaws and recommendations are given in Chapter 4.

Backing strips
If permanent backing strips are used, they must be continuous. If short lengths are welded together, full penetration butt welds must be used (see Chapter 6—Figure 54a). If the backing strip is attached by welding, unless the welds are subsequently ground out or buried in a continous weld, they must also comply with the relevant Class requirements. Since a permanent backing strip becomes an integral part of the member, it shares the applied load. Therefore, it is advisable to make it from a material of similar tensile strength to the main member to ensure that it does not yield under high applied stresses.

Edge distance
Some codes impose an edge distance criterion whereby the Class is reduced (to Class G) if a fillet weld is within 10mm of a plate edge. The criterion exists to limit the possibility of local stress concentrations occurring at unwelded edges as a result of, for example, undercut, weld spatter or accidental overwaive in manual welding. The penalty is avoided by dressing any such welds or discontinuities on the edge.

Description	Manufacturing requirements	Type No.	Class	Explanatory comments	Examples showing relevant mode of fatigue cracking, stress direction and location
Full penetration butt welds made from both sides or from one side onto a consumable insert or a temporary non-fusible backing.	Both weld beads dressed flush with any machining marks parallel to weld; weld proved free from surface breaking and significant embedded flaws by appropriate NDT.	2.1	B	Because of the extremely demanding NDT requirements, this type is only recommended for use in exceptional circumstances. Recognition of this results in the choice of Class D for dressed flush seam welds in pressure vessels.	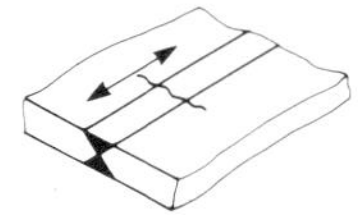 Fatigue cracks usually initiate at welding flaws
Full penetration butt welds made from both sides or from one side with or without backing. Permanent backing, which can be integral or attached by welding, must be continuous.	Automatic weld with no stop/starts and proved free from significant flaws by appropriate NDT. If permanent backing is welded, tack welds must be ground out or buried in main butt weld.	2.2	C	Accidental stop/starts are not uncommon. To meet Class C, remedial action must ensure that finished welds have surface and root bead profiles similar to those intended. Otherwise, detail down-graded to Class D.	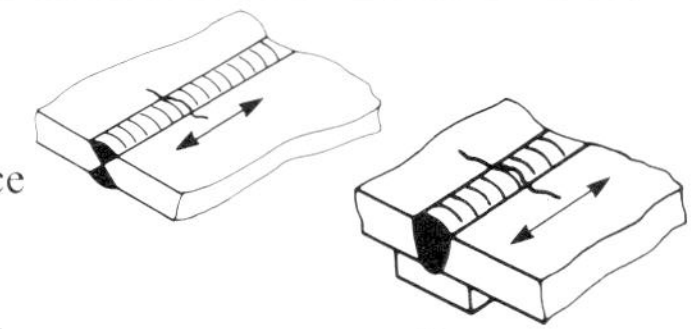
	As 2.2 but with stop/starts. Permanent backing could be attached with continuous fillet weld.	2.3	D		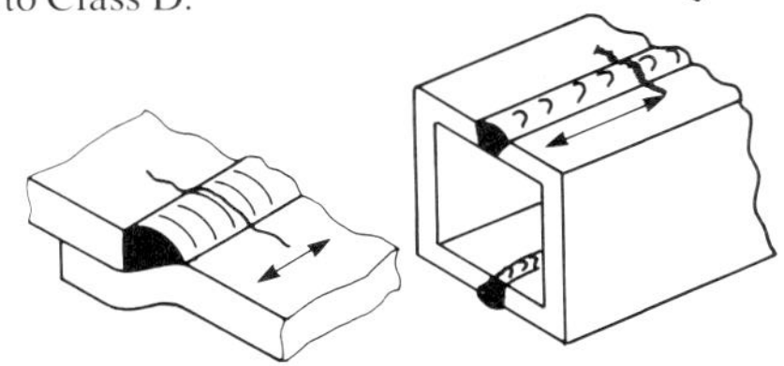
	Permanent backing strip attached with intermittent fillet welds.	2.4	E		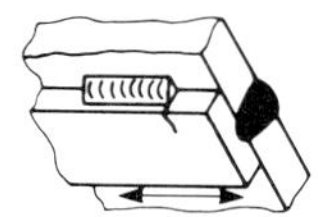

Group 3 Continuous welded attachments essentially parallel to the direction of applied stress

Potential fatigue failure modes
If the weld is continuous, fatigue cracks might initiate at the lumps on the weld surface due to stop/starts or, if these are not present, from the notch created by weld ripples. In continuous fillet welds, fatigue cracks might propagate from the weld root. A more severe stress concentration is introduced if fillet welds are discontinuous, at the weld ends, and these are the most likely sites for fatigue crack initiation.

Edge distance
See note to Group 2.

Description	Manufacturing requirements	Type No.	Class	Explanatory comments	Examples showing relevant mode of fatigue cracking, stress direction and location
Fillet and butt welded joints made from one or both sides.	Automatic weld with no stop/starts.	3.1	C	Accidental stop/starts are not uncommon. To meet Class C, remedial action must ensure that the finished weld has a surface profile similar to that intended.	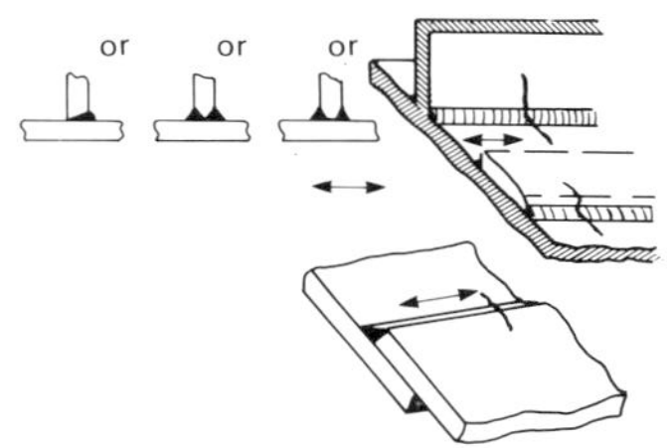
	Weld contains stop/starts.	3.2	D	For situation at ends of a flange cover plate, see detail type 8.6.	

Detail	Type	Class	Notes	
Intermittent fillet welds with:			Limiting gap ratio g/h applies even though adjacent welds may be on opposite sides of a narrow attachment (e.g. longitudinal stiffener with staggered fillet welds). Long gaps between intermittent fillet welds are not recommended as they increase the risk of corrosion and, in the case of compression members, local buckling.	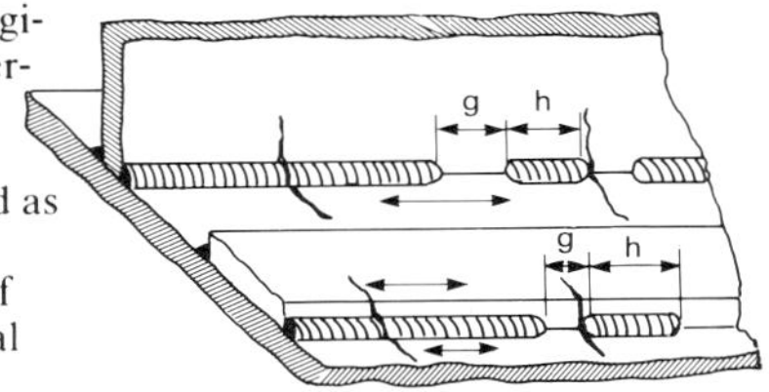
g/h ⩽2.5	3.3	E		
g/h >2.5	3.4	F		
Fillet weld at a cope hole, for weld continuing around plate ends or not.	3.5	F	Note that the existence of the cope hole is allowed for in the joint classification; it should not be regarded as an additional stress concentration in relation to fatigue cracking in the flange. For situation in web, see detail type 5.2.	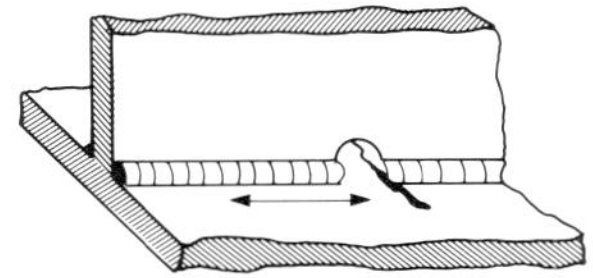

Group 4 Full penetration butt welds in plates essentially transverse to the direction of applied stress

Potential fatigue failure modes
With any weld ends dressed flush with the plate edge, fatigue cracks normally initiate at the weld toe, with the result that the fatigue life depends largely on the weld bead profile. In general, the most favourable profiles are achieved with shop welds made in the flat position, except that even in such circumstances submerged-arc welding can produce very poor profiles. If the weld overfill is dressed flush, its stress concentration effect is eliminated and embedded weld flaws usually provide the sites for fatigue crack initiation. Thus, the potentially high fatigue strength of such details is critically dependent on weld quality. As in the case of longitudinal welds (see Group 2), this potential can rarely be realised in real structures due to limitations on NDT capabilities.

Welds made from one side
There is a reduction in fatigue strength of transverse butt welds if they are made from one side only, unless a joint resembling one made from both sides can be achieved. This is possible using special consumable inserts or a suitable temporary non-fusible backing medium (e.g. copper, glass, ceramic). However, in all cases the weld should be inspected to ensure that full penetration and a satisfactory overfill shape have been achieved on the inside of the joint.

Significance of welding flaws
In all cases, it is necessary to prove that the joint is free from significant welding flaws by appropriate NDT. Fitness-for-purpose is a basis for assessing detected flaws and recommendations are given in Chapter 4.

Misalignment
Butt welds between plates are susceptible to misalignment which, under applied axial loading, has the effect of increasing the stress experienced by the joint as a result of the introduction of secondary bending stresses. Thus, the design stress must include the stress magnification effect of any misalignment (see Chapter 2). Note that this is still the case even when taper transitions in thickness are used if the centrelines of the two members are misaligned, or misalignment is blended. However, misalignment may be neglected in connections which are supported laterally, sufficiently to restrict local bending of the joint, e.g. the flanges of a girder which are supported by the web.

Size effect
For welded joints failing by fatigue crack growth from the weld toe, there is a size effect such that fatigue strength can decrease with increase in plate thickness (see Chapter 2). In such cases, a plate thickness design penalty is introduced whereby the allowable design stress for the particular classification is reduced by the factor $(B_{ref}/B)^{0.25}$, where B = plate thickness and B_{ref} is a reference thickness to which the design S–N curves are directly applicable. For plate welded joints, the reference thickness normally used is 22mm, but other values may be used for other cases.

Description	Manufacturing requirements	Type No.	Class	Explanatory comments	Examples showing relevant mode of fatigue cracking, stress direction and location
Full penetration butt weld joining plates of equal width and thickness made from both sides or from one side onto consumable insert or temporary non-fusible backing.	Weld overfill dressed flush and welds proved free from surface breaking and significant embedded flaws by appropriate NDT. Any misalignment blended with slope 51 in 4.	4.1	C	Because of extremely demanding NDT requirements, this type is only recommended for use in exceptional circumstances and certainly not in structural work. Recognition of this results in choice of Class D for dressed flush seams in pressure vessels.	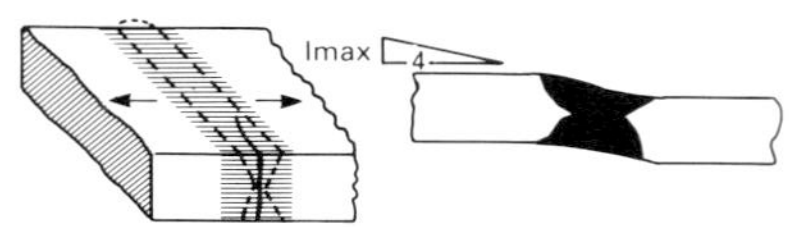 Fatique cracks usually initiate at weld flaws

Full penetration butt weld joining plates of any width or thickness, with changes tapered to ⩽ 1 in 4 slope, except that up to 15% thickness change can be accommodated in the weld profile without taper.	Shop welds made from both sides in the flat position either manually or by automatic process other than submerged-arc.	4.2	D	Welds made positionally, on-site or by submerged-arc process tend to have poor overfill shapes. Hence they are down-graded from D to E unless a favourable weld profile can be produced. Welds designed to Class D which fail to meet the criteria stated should be dressed flush in the regions affected. Plate thickness design penalty applicable.	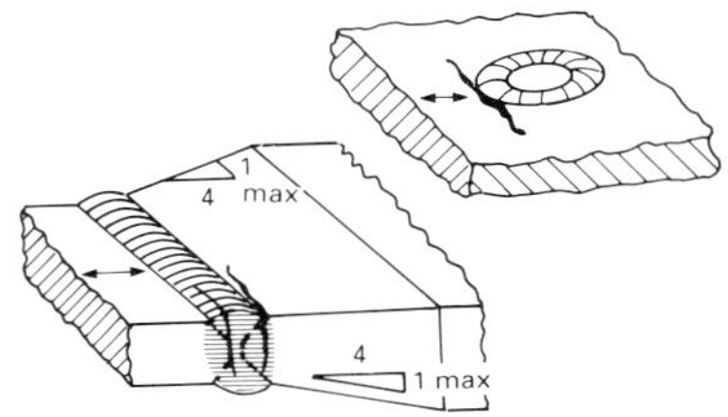 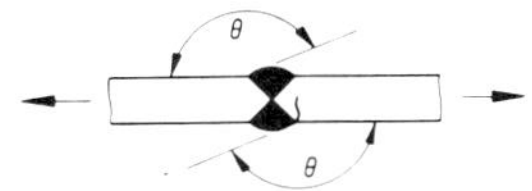
	Welds made by any means from both sides or from one side onto consumable insert or temporary non-fusible backing. Overfill profile $\theta \geqslant 150°$ Overfill profile $\theta < 150°$	 4.3 4.4	 D E		
	In welds made from one side, full penetration assured and in all cases welds proved free from significant flaws by appropriate NDT.	4.5		Not recommended for fatigue loaded joints, since fatigue life critically dependent on root condition. If full penetration can be assured, then Class E.	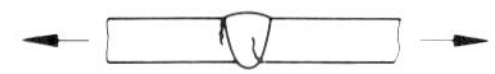
	Weld made from one side onto permanent backing which is either integral or attached by tack or fillet welds.	4.6	F		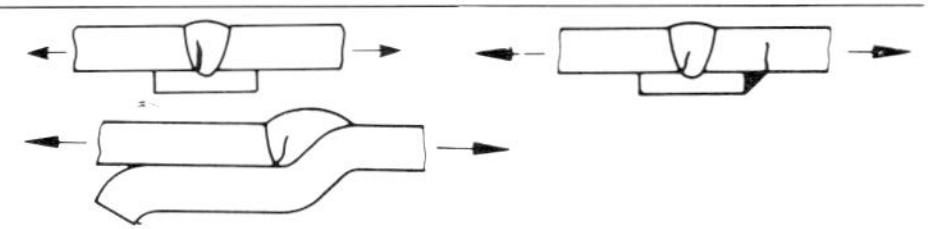
Full penetration butt weld joining plates of unequal width which are not tapered.	Weld ends ground to radius $\geqslant 1.25 \times$ thickness.	4.7	F_2	Stress concentration has been taken into account in the joint classification. Step changes in width can often be avoided by the use of shaped transition plates, arranged so as to enable butt welds to be made between plates of equal width (see sketch). Plate thickness design penalty applicable.	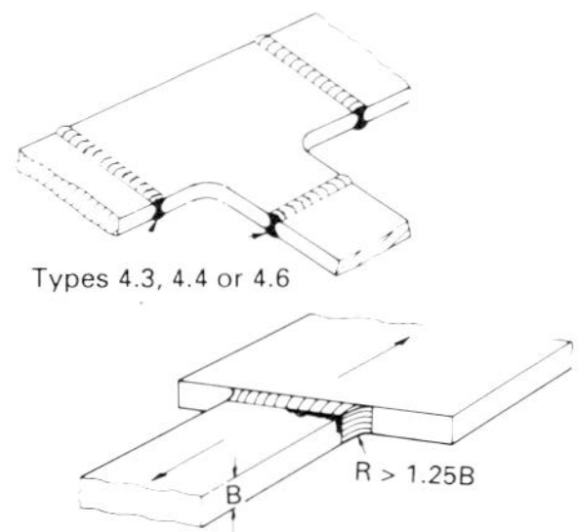

Group 5 Full penetration butt welds in sections with weld essentially transverse to the direction of applied stress

Potential fatigue failure modes
Butt welds between sections are prone to welding flaws, e.g. in region of the web/flange junction. They are difficult to detect and may provide the sites for fatigue crack initiation. As a result, the classification for butt welds in sections may be lower than those for joints in simple plates. However, special preparations, procedures and inspections may be undertaken in exceptional circumstances and the appropriate classification for Group 4 details assumed.

In some cases, the weld toe will still provide the most likely site for fatigue crack initiation, except in welds made from one side when the root will be more critical.

Significance of welding flaws
In some cases, it is necessary to prove that the joint is free from significant welding flaws by appropriate NDT. Fitness-for-purpose is a basis for assessing detected flaws and recommendations are given in Chapter 4.

Size effect
For cases which refer to weld toe fatigue cracking, the plate thickness design penalty $(B_{ref}/B)^{0.25}$, as described in notes to Group 4, is applicable.

Description	Manufacturing requirements	Type No.	Class	Explanatory comments	Examples showing relevant mode of fatigue cracking, stress direction and location
Full penetration butt welds between rolled, extruded or built-up sections.	Butt welds between sections are prone to weld defects, which are difficult to detect, in the region of the web/ flange junction. Special preparations, procedures and inspection may be undertaken in exceptional circumstances and types 4.4 or 4.6 as appropriate may be assumed. Dressing of the weld overfill is advised to overcome poor	5.1	F_2		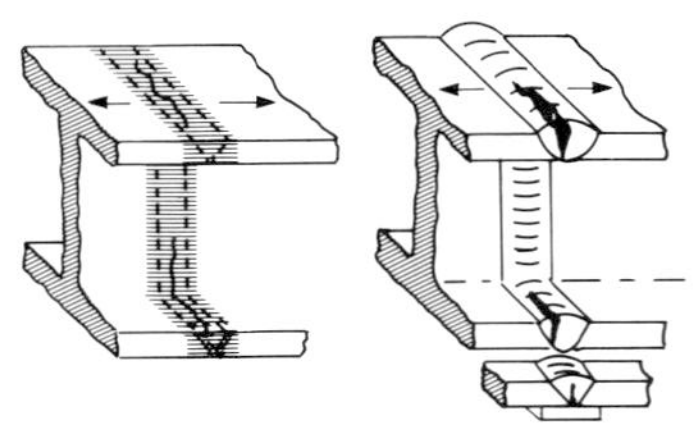

	weld bead shape resulting from the greater misalignments which may occur in the jointing of sections.				
	Semi-circular cope hole used, e.g. to improve access to flange butt weld when webs or longitudinal stiffeners have already been attached. Weld end and weld overfill over distance ⩾ r dressed flush.	5.2	D	Class refers to web butt weld where it meets cope hole and is based on stress range in web adjacent to cope hole multiplied by SCF of 2.4. Cope holes of 45° mitre are not recommended.	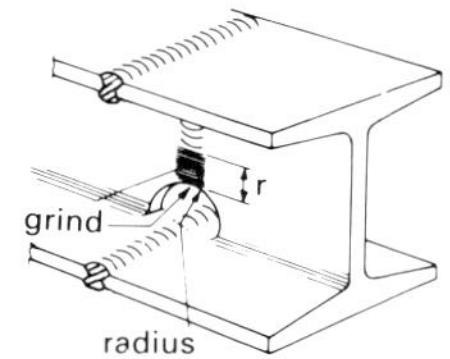
Full penetration butt welds between hollow sections, including joints between circular section tubes and conical sections.	Welds made from both sides with overfill dressed flush and weld proved free from surface breaking and significant embedded flaws by appropriate NDT.	5.3	C	Because of extremely demanding NDT requirements, this type is only recommended for use in exceptional circumstances and certainly not in structural work.	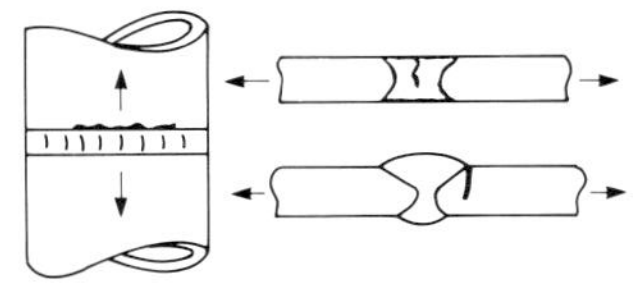
	Welds made from both sides or from one side onto consumable insert or temporary non-fusible backing.	5.4	E	Plate thickness design penalty applicable.	
	Welds made from one side onto permanent backing which is either integral or attached with tack or fillet welds.	5.5	F	Beware of corrosion on backing strip if this is attached sometime before making butt weld (e.g. when used as an alignment aid during fabrication).	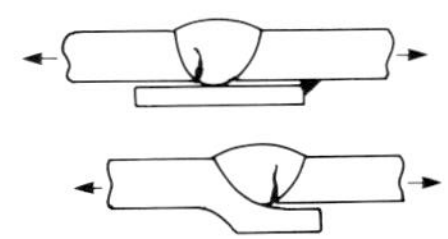
	Welds made from one side without backing in which full penetration is assured.	5.6	F_2	Plate thickness design penalty applicable.	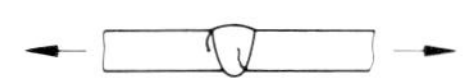

Group 6 Fillet and T butt welds at end connections between loaded members

Potential fatigue failure modes
Fatigue cracks in cruciform or T joints with full penetration welds will normally initiate at the weld toe and propagate through the parent plate. In joints made with load-carrying fillet or partial penetration butt welds, cracking may initiate either at the weld toe or at the weld root, the latter propagating through the weld throat. In welds parallel to the direction of the applied stress, however, weld failure is uncommon, although its possibility should be checked. Cracks normally initiate at the weld end and propagate into the plate perpendicular to the direction of applied stress.

Edge distance
See note to Group 2.

Size effect
For cases which refer to weld toe fatigue cracking, the plate thickness design penalty $(B_{ref}/B)^{0.25}$ is applicable (see notes to Group 4). This penalty is not applicable for cases of fatigue cracking from the weld root through the weld throat or fatigue cracking from the weld end on the edge of the stressed plate.

Description	Manufacturing requirements	Type No.	Class	Explanatory comments	Examples showing relevant mode of fatigue cracking, stress direction and location
Cruciform or T joints between plates, sections or built-up members.	Full penetration butt weld.	6.1	F	Plate thickness design penalty applicable.	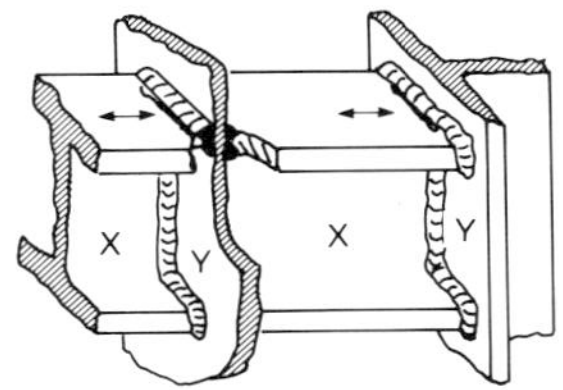
	Partial penetration butt or fillet weld.	6.2	F_2	Plate thickness design penalty applicable.	
				In this type of joint, failure is likely to occur in the weld throat unless the weld is made sufficiently large (see joint type 6.3).	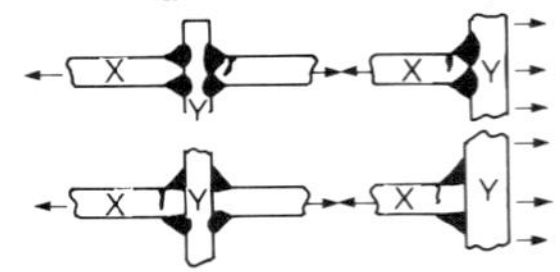

		6.3	W	Based on stress range on weld throat area, assuming that none of the load is carried in bearing between the plates.	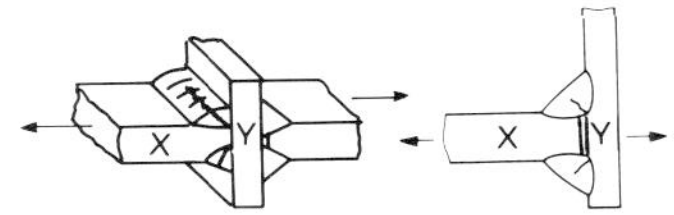
Fillet welded lap joints between plates or sections.	Lap joint on both surfaces of member X symmetrically.	6.4	F_2	Where narrow attachment transfers the entire load out of a wider member, assume that all load in that member acts over effective width shown.	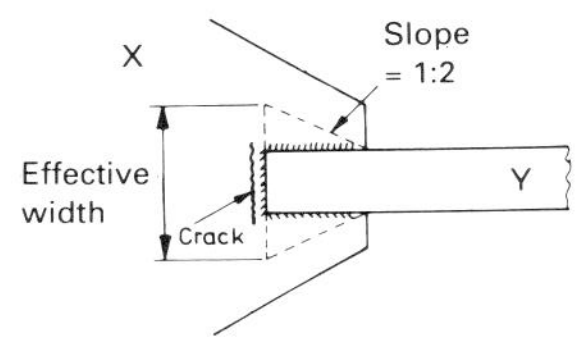
	Lap joint on only one surface of member X.	6.5	G	Plate thickness design penalty applicable. For failure in attachment plate, see joint type 6.7.	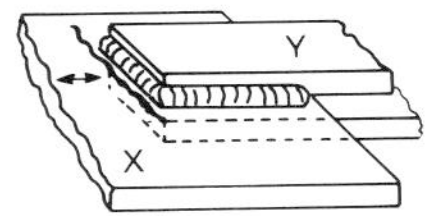
	Any lap joint.	6.6	W	Based on stress range on weld throat area.	
	Any lap joint. Grind any undercut and avoid weld returns around laps.	6.7	G	This type applies regardless of shape of end of attachment. In all cases, care should be taken to avoid undercut on corners or to grind it out to a smooth profile should it occur. In particular, weld returns across a corner should be avoided and the use of the attachment plate wider than member X is not recommended.	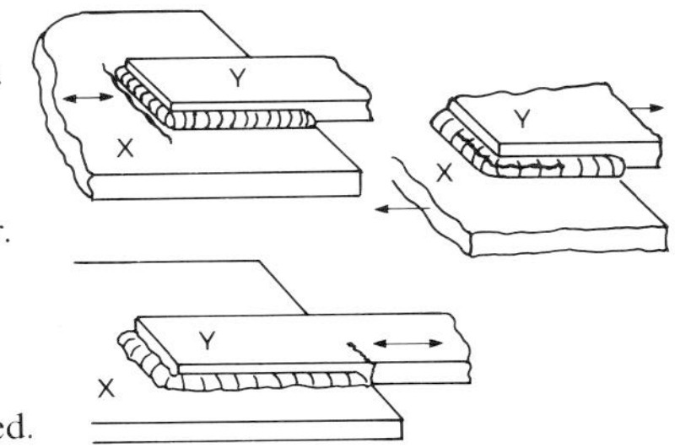

Group 7 Slotted connections and penetrations through stressed members

Description	Manufacturing requirements	Type No.	Class	Explanatory comments	Examples, showing relevant mode of fatigue cracking, stress direction and location
Slotted through member.	Full penetration butt weld with length of member Y parallel to applied stress ⩽ 160 mm. > 160 mm.	 7.1 7.2	 F F_2	Plate thickness design penalty applicable. Classifications do not apply to fillet welded joints (see joint type 6.2). However, they do apply to loading in either L or T direction.	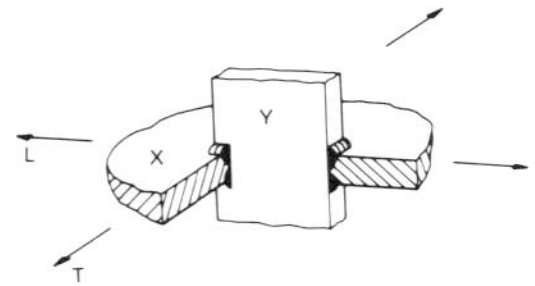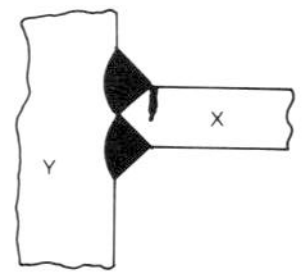
Penetrations, including tubes (e.g. branch connections in pressure vessels).	Fillet or butt welded.	7.3	F	The relevant stress range should include the SCF due to the structural discontinuity introduced by the penetration. Plate thickness design penalty applicable.	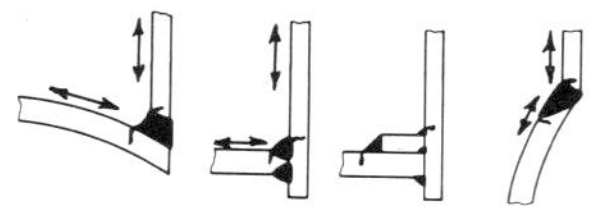

Full penetration butt weld.	7.4	D	Refers to failure in weld metal from surface or embedded discontinuity; based on stress range on weld cross-section, multiplied by SCF due to structural geometry.	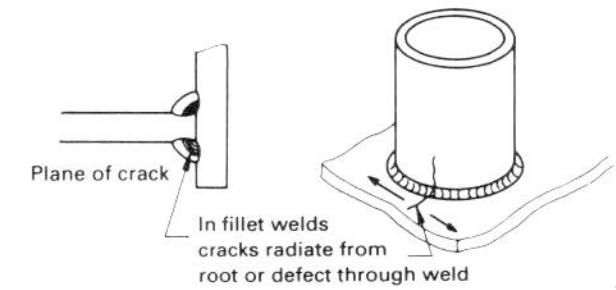
Partial penetration butt or fillet weld.	7.5	F		
	7.6	W	Refers to fatigue failure across weld throat; based on stress range on weld throat area.	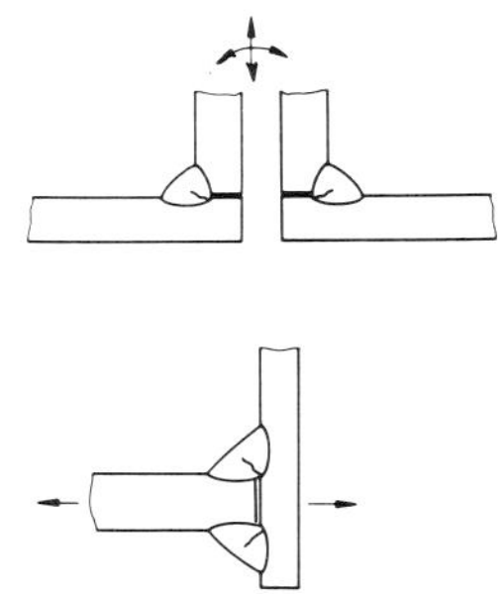

Group 8 Welded attachments on the surface or edge of a stressed member

Potential fatigue failure modes
When the weld is parallel to the direction of the applied stress, fatigue cracks normally initiate at the weld ends and propagate into the parent plate. When the weld is transverse to the direction of stressing, they usually initiate at the weld toe. For attachments involving a weld on one side only, fatigue cracks might also initiate at the weld root and propagate into the parent plate.

Edge distance
The stress concentration due to an attachment is increased, and the fatigue strength is therefore reduced, if the weld end is located on the edge of the stressed member rather than on its surface. This is the reason for the edge distance criterion, which may apply to the plate surface attachments, as discussed in notes to Group 2.

Size effect
The decrease in fatigue strength with increase in attachment size is because more load is transferred into the larger attachment, giving an increase in stress concentration at the weld toe.

In the case of attachments to the plate surface, the plate thickness design penalty $(B_{ref}/B)^{0.25}$ is applicable. The penalty is not applicable to edge attachments.

Description	Manufacturing requirements	Type No.	Class	Explanatory comments	Examples showing relevant mode of fatigue cracking, stress direction and location
Attachment of any shape with edge in contact with stressed member lying essentially transverse to the direction of stressing.	Fillet or butt welds, with welds continuous around ends or not.				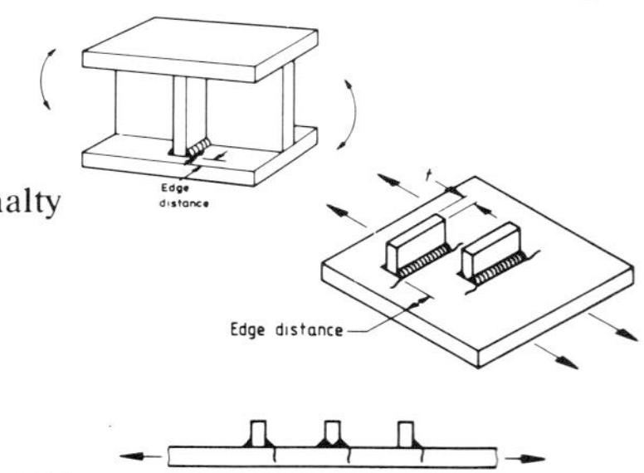
	$t \leqslant 55$ mm.	8.1	F	Plate thickness design penalty applicable.	
	$t > 55$ mm.	8.2	F_2		
	Butt welds should be made with an additional reinforcing fillet to produce similar toe profile to that which would exist in a fillet welded joint.				

Attachments to girder webs in regions of combined bending and shear.		8.3	E	Principal stress range adjacent to weld used. Plate thickness design penalty applicable, but unlikely to be relevant.	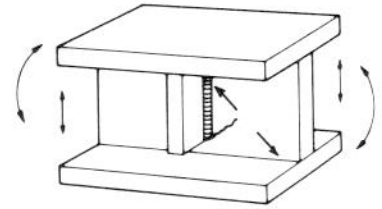
Attachments of any shape with surface in contact with stressed member.	Fillet weld continuous around ends or not:			Plate thickness design penalty applicable.	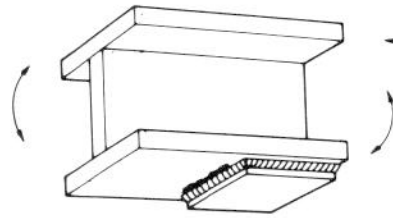 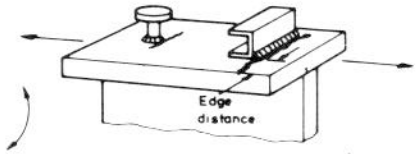
	L ⩽ 160 mm, W ⩽ 55mm.	8.4	F		
	L > 160 mm, W ⩽ 55 mm.	8.5	F_2		
	L > 160 mm, W > 55 mm.	8.6	G	Classification includes stress concentration arising from normal eccentricities in the thickness direction.	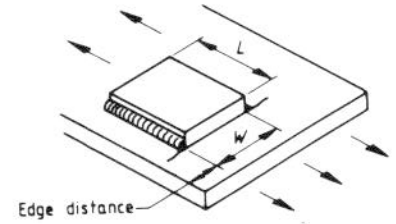
	Cover plate of any length and wider than girder flange.	8.7	G	Not recommended due to likelihood of undercutting flange edge. Weld returns across corner should be avoided. Plate thickness design penalty applicable.	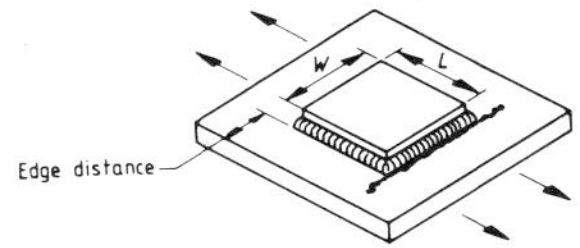
Attachments of any shape on or close to the edge of a stressed member.	Any size of butt or fillet welded attachment with weld continuous around ends or not.	8.8	G	Avoid undercut on corners or grind it out to a smooth profile should it occur. Weld returns across a corner should be avoided. Note that the classification applies to all sizes of attachments. It would therefore include for example the junction of two flanges at right angles. In such situations a low fatigue classification can often be avoided by the use of a transition plate (see also joint type 4.7).	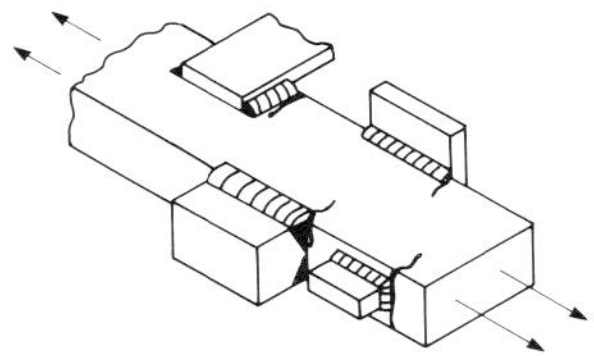

Group 8 cont'd

Description	Manufacturing requirements	Type No.	Class	Explanatory comments	Examples showing relevant mode of fatigue cracking, stress direction and location
Attachment of any shape with edge in contact with stressed member lying essentially parallel to the direction of stressing.	Fillet or butt welds, with welds continuous around ends or not. $L \leqslant 160$ mm, $W \leqslant 55$ mm. $L > 160$ mm, $W \leqslant 55$ mm. $L > 160$ mm, $W > 55$ mm.	 8.9 8.10 8.11	 F F_2 G	Plate thickness design penalty applicable.	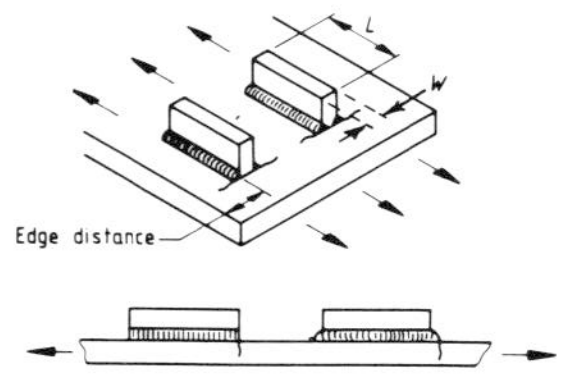
		8.12	F	Note that the design stress must include any local bending stress adjacents to the weld end. Plate thickness design penalty applicable.	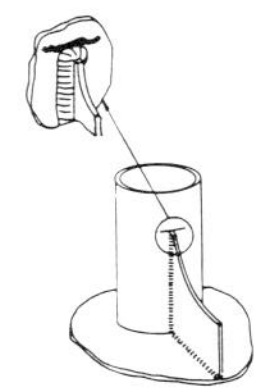

The classifications given in this Table are based on those originally published by T R Gurney ('Fatigue design rules for welded steel joints', The Welding Institute Research Bulletin, Vol. 17, May 1976). They broadly comply with those subsequently included in the UK bridge design rules (BS 5400 Pt. 10, see Bibliography) and in other UK codes which have adapted the BS 5400 rules for application to other structures (see Bibliography). They are also similar to the international rules in Eurocode No. 3 and those produced by the International Institute of Welding (see Bibliography). However, designers using the Table who are required to meet a particular specification should ensure that the classification and conditions given meet that specification.

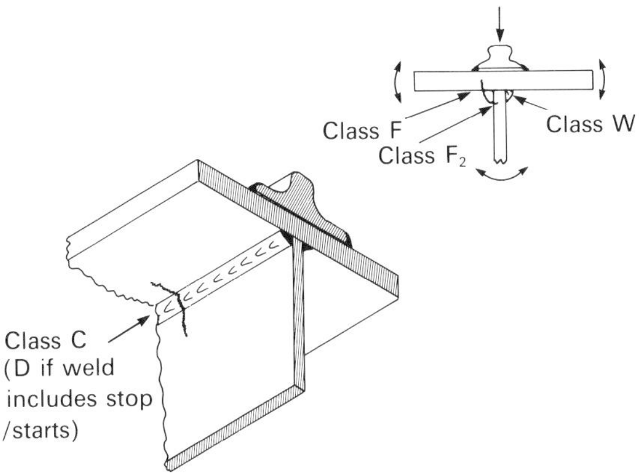

2 Classifications for web-to-flange fillet weld in crane girder.

two sites, and the weld itself, although the fatigue strength, and hence class, for weld failure would depend upon where the fatigue crack initiated, e.g. weld root in transverse welds, weld ripples or stop/start positions in longitudinal welds, weld defects. To illustrate this consideration, Fig. 2 shows the various classifications for a single joint depending upon the direction of loading and potential site for fatigue cracking.

Knowledge of potential sites for fatigue cracking is also necessary if it is required to specify in-service inspection procedures. Such measures may be considered necessary in critical members whose failure would be dangerous or expensive to repair or replace.

It is perhaps worth emphasising the fact that the classifications refer to the fatigue performance of welded joints and not to their quality. A Class G weld is not necessarily of lower quality than one of Class E; it simply has a lower fatigue strength which, in practice, may be quite adequate. There is no reason to suppose that the welder, or even the inspector unless instructed, will recognise the classifications. Where necessary, the designer should ensure that conditions relating to welding method, weld quality or inspection associated with the choice of classification are clearly described to the fabricator.

Unclassified joints

Although the classification scheme includes most of the commonly used welded joints, it is recognised that many are not included. This shortcoming arises partly because it is confined to cases which have been included in Standards. Other fatigue test data are available, for example for electron beam butt welds, and ultimately they will find their way into design rules. However, in other cases there is simply a lack of available test data. In particular, there are few data for welds made using processes other than manual metal arc, submerged-arc and gas shielded welding. A third problem is that although test data may be available, there is no suitable means of interpreting them in a general way for design. A notable example is resistance spot welds between thin sheets, where a suitable stress to relate the fatigue behaviour of simple spot welded specimens to the behaviour of real spot welded structures is still sought.

In general, the classification of a detail not presently classified could be established by reference to published work or by carrying out special tests. The resulting test data should be analysed to establish an S–N curve which embodies the same probability of survival as the design S–N curves, taking due account of the statistical significance of the number of test data available, and compared with the grid of design S–N curves. The classification would then be that corresponding to the design S–N curve next below the deduced S–N curve.

Design S–N curves

For each class of joint the general equation relating applied stress range, S and fatigue life, N, is

$$S^m N = a\Delta^d \qquad [1]$$

where m is the slope, a is a constant which refers to the mean S–N curve, d is the number of standard deviations of log N below the mean, and Δ depends on the standard deviation of log N, σ, such that $\sigma = \log(1/\Delta)$. For a particular design S–N curve, the equation may be written

$$S^m N = A \qquad [2]$$

where $A = a\Delta^d$. For the mean S–N curve, A = a; for the M–1SD, $A = a\Delta$; for the M–2SD, $A = a\Delta^2$, etc. Values of m and Δ together with values of A for the mean (also equal to a), M–1, M–2SD, S–N curves are given in Table 2 (for as-welded joints); equations for other confidence limits can be calculated easily by putting the required value of d into [1].

The M and M–2SD S–N curves are shown graphically in Fig. 3. Since the curves are linear when plotted in logarithmic terms they may be extrapolated linearly to higher or lower stresses as required, although values should not be taken from the curves but rather should be calculated using the equations.

In practice the S–N curves for constant amplitude loading would cut off at the endurance limit, S_0, which, for all classes in air, is assumed to correspond to an endurance of 10^7 cycles. Values of S_0 are included in

Table 2. A cut off at high stresses is also implied by general design rules in the sense that the permissible static stress for the steel concerned should not be exceeded, i.e. by the maximum stress developed under fatigue loading. However, this limit may not always apply and, in principle, the S–N curve can be extrapolated beyond yield, into the low-cycle fatigue regime. A condition is that the overall structural behaviour remains elastic, as discussed further in Chapter 2.

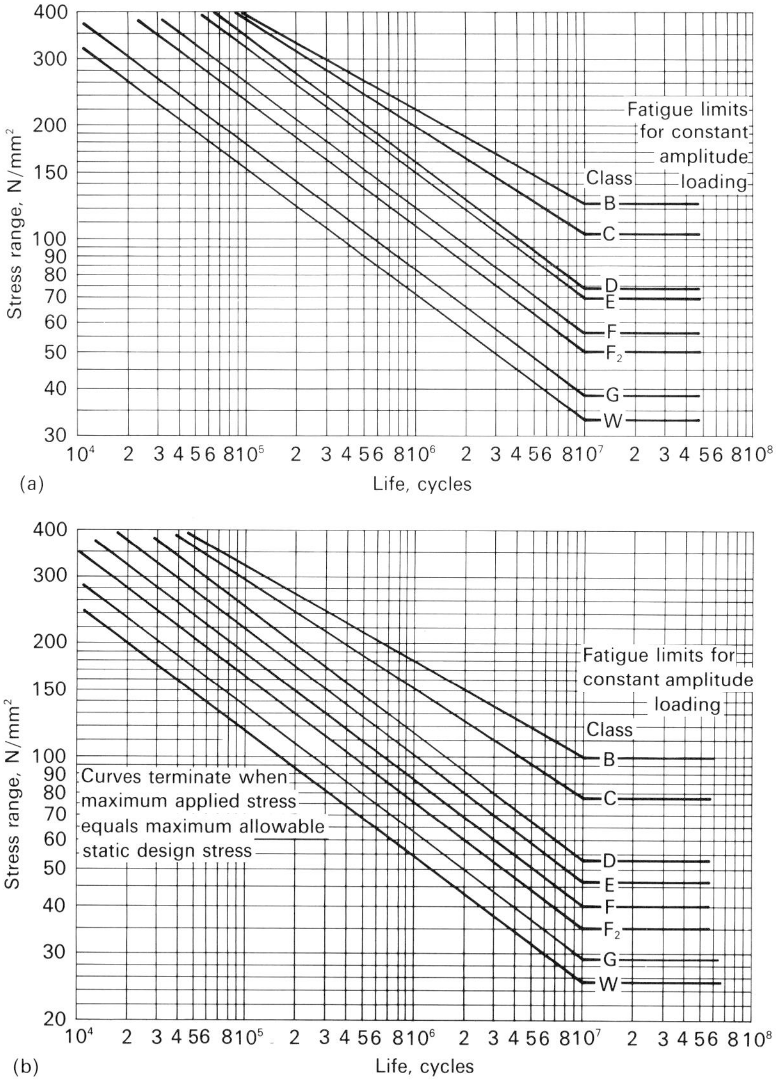

3 Standard S–N curves: (a) Mean lines (50% probability of failure); (b) M–2SD lines (2.3% probability of failure).

Table 2 Details of design S–N curves (for as-welded steel joints)

Curve	m	Δ	A	S_0 N/mm²
Class B				
Mean			2.34×10^{15}	124
Mean−1SD	4.0	0.657	1.54×10^{15}	111
Mean−2SD			1.01×10^{15}	100
Class C				
Mean			1.08×10^{14}	102
Mean−1SD	3.5	0.625	6.75×10^{13}	89
Mean−2SD			4.22×10^{13}	78
Class D				
Mean			3.99×10^{12}	74
Mean−1SD	3.0	0.617	2.46×10^{12}	63
Mean−2SD			1.52×10^{12}	53
Class E				
Mean			3.29×10^{12}	69
Mean−1SD	3.0	0.561	1.85×10^{12}	57
Mean−2SD			1.04×10^{12}	47
Class F				
Mean			1.73×10^{12}	56
Mean−1SD	3.0	0.605	1.05×10^{12}	47
Mean−2SD			6.33×10^{11}	40
Class F_2				
Mean			1.23×10^{12}	50
Mean−1SD	3.0	0.592	7.28×10^{11}	42
Mean−2SD			4.31×10^{11}	35
Class G				
Mean			5.70×10^{11}	38
Mean−1SD	3.0	0.662	3.77×10^{11}	33
Mean−2SD			2.50×10^{11}	29
Class W				
Mean			3.70×10^{11}	33
Mean−1SD	3.0	0.654	2.42×10^{11}	29
Mean−2SD			1.58×10^{11}	25

Effect of welded joint dimensions

The size effect observed in the fatigue behaviour of welded joints was discussed in Pt 1, where particular attention was drawn to the reduction in fatigue strength which can arise with increases in attachment size or main plate thickness for details failing from the weld toe.

The effect of attachment size is introduced into the design rules as part of the classification system. Thus, referring to Table 1, the classification for a stressed member with surface attachments varies from F to G depending on the attachment length or width.

The effect of plate thickness, on the other hand, is introduced as a factor by which the permissible design stress range must be multiplied, namely $(\frac{22}{B})^{0.25}$, where B = plate thickness in millimetres. Weld details to which this thickness correction must be applied are indicated in Table 1. The factor was deduced from fatigue test results obtained from members containing transverse fillet or butt welded joints, as illustrated in Fig. 4.

Fracture mechanics analysis of the fatigue behaviour of welded joints failing from the weld toe provides some insight into the reason for the size effects mentioned and draws attention to factors which could reduce the design penalties presently assumed. Attachment size and plate thickness are important because they influence the stress concentration of the weld toe. This stress concentration affects the stress intensity factor, K, and hence the rate of crack propagation da/dN, of a crack while the crack is in its zone of influence (for further information about the corresponding function M_k, see Appendix III). This zone is normally a function of plate thickness; for example, for the cases illustrated in Fig. 4, in which the 'attachment' lengths (L see Fig. 5) are similar to the plate thickness (B), the stress concentration

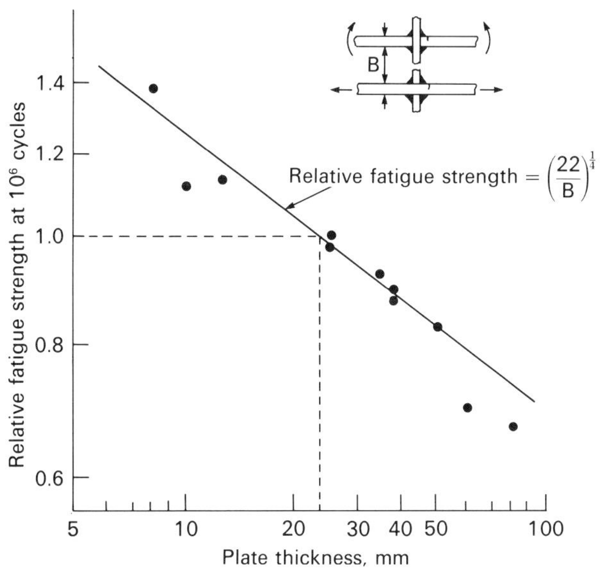

4 Effect of plate thickness on fatigue strength (R ~ 0).

influences K until the crack is approximately ⅓ of the way through the thickness. Thus, since the depths of inherent weld toe flaws from which fatigue cracks propagate (see Pt 1) are not dependent upon plate thickness, da/dN is affected by the stress concentration to a greater extent for thick plates than for thin, and the 'thickness effect' seen in Fig. 4 results.

However, there is an important competing effect which is not apparent from the results in Fig. 4. The *magnitude* of the stress concentration of the weld toe is found to depend on the proportions of the joint, L/B, such that a small attachment to a thick plate may result in a lower stress concentration factor than a long attachment to a thin one, and hence no adverse effect of increased thickness. In contrast, the data in Fig. 4 refer to joints in which the attachment size was scaled up with main plate thickness (ie L/B~ constant); an even greater 'thickness effect' would be expected if L/B increased with increase in B.

Ideally, account should be taken of both L and B, when allowing for the size effect in design.

A parametric study using fracture mechanics has indicated ways of accounting for L/B when calculating the stress range, to give a stress analogous to the 'hot spot' stress used in tubular joint design (see Chapter 2), which can then be used simply in conjunction with the design S–N curve for transverse butt welds, on the basis that all stress

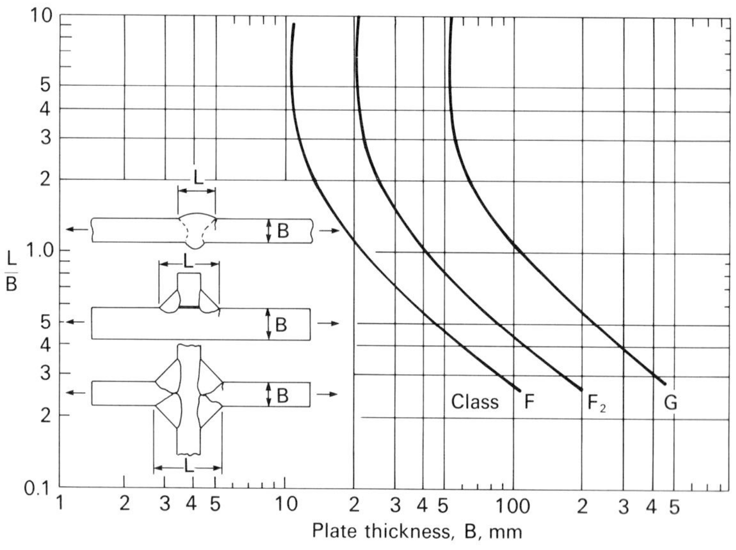

5 Definition of attachment length and joint dimensions corresponding to design classes F, F_2 and G calculated using fracture mechanics.

concentration effects except that due to the local weld toe have been incorporated. Alternatively, the combination of dimensions L and B corresponding to the present design classes for attachment details can be presented graphically, Fig. 5. Further experimental work is needed to validate these theoretical indications, which could provide a basis for relaxation of the present 'thickness correction' design penalty.

It will be apparent that the weld toe stress concentration has a crucial influence on the 'size effects' discussed. In view of this, relaxation of the design penalty should also be possible if either the magnitude or extent of the stress concentration can be reduced. Thus, the 'size effect' may prove to be less significant in the higher design class details, and in joints in which the weld profile has been improved, for example by grinding (see Chapter 5). Further research is required to confirm these possibilities.

Chapter 2

2 Stresses used with the fatigue design rules

Nominal stress

The underlying principle upon which the fatigue design rules are based is that the fatigue strength of a weld detail in a structure can be estimated on the basis of the performance of that same detail in a simple laboratory specimen. For example, considering fatigue failure from the weld toe, the ends of the gussets in Fig. 6 would be assumed to be represented by the ends of the attachment in the test specimen. The test specimen incorporates the local stress concentration caused by the geometry of the weld and local features of the gusset, and so the logical reference stress under which the fatigue strengths of the two details should be similar is the nominal stress in the plate near the weld detail.

In the design rules the joint classifications take account of the influence of the local stress concentrations introduced by the weld details themselves, as illustrated in the sketches in Table 1. The design stresses quoted should be regarded as nominal stress ranges in the vicinity of the detail under consideration relevant to the particular mode of fatigue cracking being considered.

Thus, in evaluating the possibility of fatigue cracking from the toe of a weld, the nominal stress in the parent material at the weld toe, calculated without taking into account any increase in stress because of the weld profile, should be used. In the examples shown in Fig. 7 the nominal stresses at the butt or fillet weld toes (sections A–A and B–B) are assumed to be the same as they would be in the absence of the welds, such as at section C–C, and the relevant stress for design purposes is S_1.

For longitudinally loaded continuous joints, contained in Groups 2, 3 and Table 1, the relevant stress is that adjacent to the weld acting parallel to it. If, on the other hand, the likelihood of cracking through the throats of load-carrying welds is to be considered (Class W), the relevant stress is the stress range across the weld throat, calculated on the load range carried by the weld divided by the minimum weld throat area.

Combined stresses

Most fatigue data have been obtained under unidirectional, axial or bending loads, for which the relevant nominal stress range to express the fatigue strength of the detail can be easily defined. For example, for a transverse butt weld in an axially loaded plate it would be:

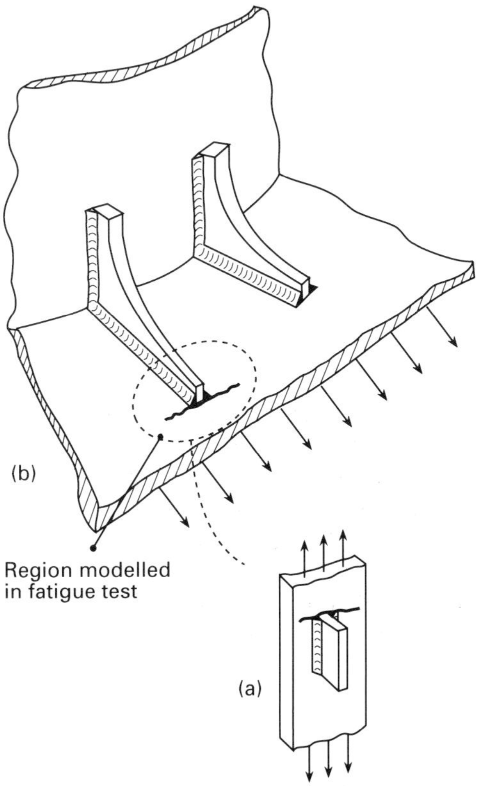

6 Relationship between: (a) Fatigue test specimen; (b) Welded structure.

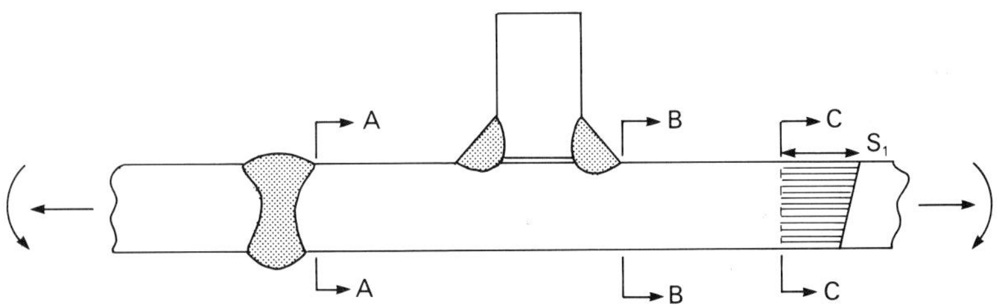

7 Illustration of nominal stress.

$$\frac{\text{maximum load - minimum load}}{\text{cross-sectional area of plate at weld toe}}$$

However welded joints in real structures may be subjected to a combination of types and directions of loading. In these circumstances the stresses in parent material to be used for fatigue assessment are the principal stresses. Thus, for example, in the design of girder webs with stiffeners (see detail 8.3 in Table 1) the combined effect of bending and shear must be considered.

In general terms, the maximum principal stress range is used, calculated as the greatest algebraic difference between principal stresses occurring during the whole loading event. This approach is probably reasonable if the principal planes are less than 45° apart, but it may be too conservative for greater angles. In practice, such situations can arise, for example at the web stiffener if the girder carries a rolling load. An alternative approach is to calculate the maximum principal stress range from the maximum changes in the stress components during the loading event. Such stress changes would be obtained directly from a strain gauge rosette fixed near the weld detail under consideration, in, for example, a prototype. Even when using principal stresses, account should be taken of the welded joint orientation with respect to the stress directions. In particular, the maximum principal stress range may not act in a direction which corresponds to the lowest classification for the joint. In this respect, a continuous weld (e.g. a seam in a vessel) would be classified as longitudinal with respect to stresses acting within 30° of the weld axis, and transverse for other directions.

For load-carrying welds, the stress in the weld should be the vector difference between the maximum and minimum vector sum of the stresses on the weld throat.

Welds in regions of stress concentration

The stress near a weld detail may be increased above the nominal stress in the member as a result of openings in the member or the structural arrangement of the joint, as well as the local stress concentration stemming from the weld detail. This is actually the situation with **attachments** to stressed members, but here the classification system given in Table 1 incorporates the additional stress concentration by relating the joint class to the attachment size, as discussed above. Similarly, the effect of having a fillet weld end at a cope hole in the web of an I beam (Type 3.5 details) is covered by the classification system. However, for cases not so covered, if a weld falls within the zone of influence of a structural stress concentration, this must be taken into account when calculating the effective stress range to be used in conjunction with the design rules.

Welds adjacent to openings

A simple example, that of a transverse fillet weld meeting an opening in a stressed plate, was described in Pt 1 but is briefly reiterated here, Fig. 8. Nominal stress, based on net area, is S and the weld detail is rated as Class F. Where, however, the weld meets the opening, the stress in the plate is K_tS, K_t being the elastic stress concentration factor associated with the opening, and it follows that the detail must be designed so that K_tS represents an acceptable stress range for the life required. A similar consideration arises if, in addition to having fillet weld ends at a cope hole in a beam, accounted for in classification, a butt weld splice joint coincides with the cope hole as shown in Fig. 9; a value for K_t at the position of the weld must be introduced and used in conjunction with the appropriate butt weld class (note that in these circumstances it would be necessary to grind smooth the ends of the welds at the edge of the opening, see Type 5.2).

A variation of the influence of an opening in a stressed member arises when an inserted boss or tube is attached by a peripheral weld, Fig. 10a. In this instance the maximum stress concentrations from the opening, at point X, do not coincide with transverse weld toes but with regions which correspond to the condition of continuous longitudinal

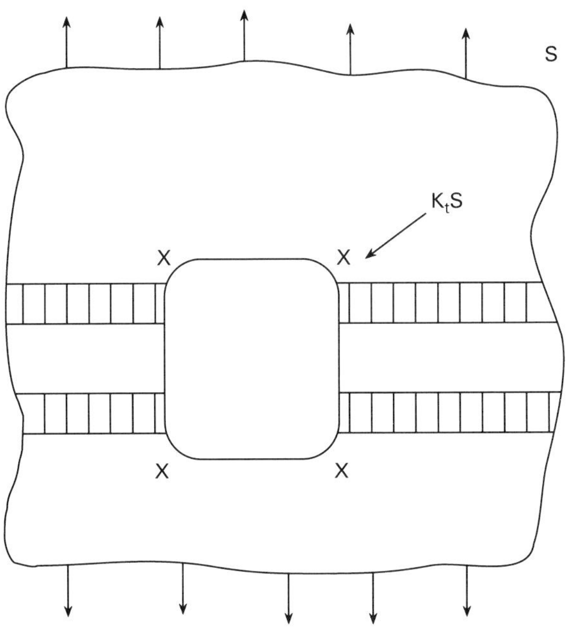

8 Fillet weld adjacent to opening in stressed plate.

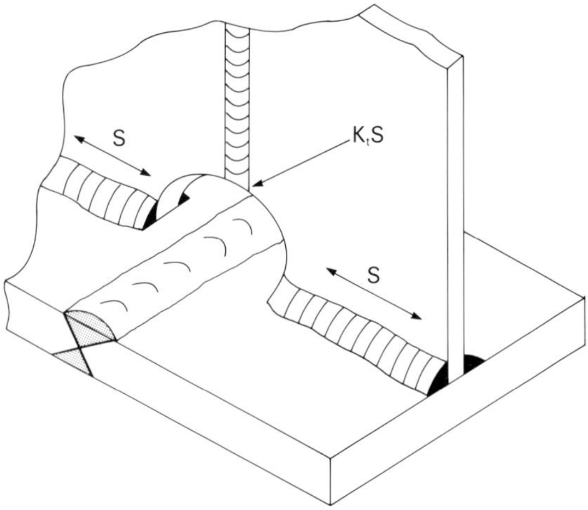

9 Butt weld in region of stress concentration at edges of cope hole.

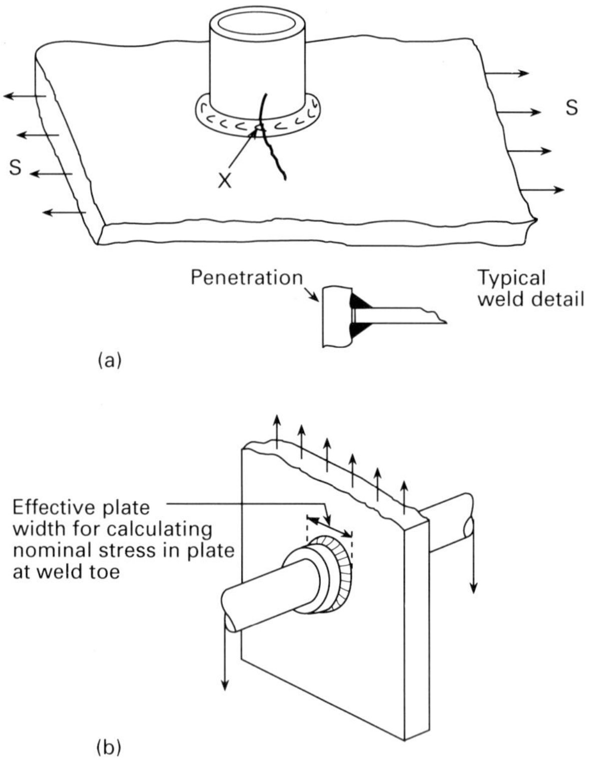

10 Stress concentrations associated with penetrations.

welds. Again, the effective stress range would be taken as K_tS and the use of Class D would be advised, for fillet or butt welds, by virtue of the difficulty of avoiding root defects in details of this kind. Cracking from the toe or root of the weld in areas transverse to the applied stress would still need to be considered, in particular when the attached member itself is loaded, as shown, for example, in Fig. 10b. Stress range at the weld toe in the plate is calculated on the assumption that all the load in the plate is carried only by the plate section in the vicinity of the weld (as illustrated); the principle is the same as that shown for detail 6.4 in Table 1.

Nozzles in pressure vessels

Nozzles in pressure vessels provide a major example of welds located at openings and the classifications given in Table 1 are used in conjunction with the nominal stress range in the vessel shell or the nozzle itself, depending on the detail being assessed, multiplied by the appropriate stress concentration factor.

In such cases, the resulting local stress may exceed the yield, but as long as the surrounding material remains elastic, the overall behaviour of the structure will also be elastic (i.e. strain cycling in the region of the nozzle will be under 'strain controlled' conditions). The design S-N curve is still applicable in such circumstances, expressed in terms of the pseudo-elastic stress range, calculated simply as strain range × elastic modulus. Figure 11, which shows both high-cycle and low-cycle (strain controlled) fatigue test data obtained from fillet welds, confirms this correlation between the high- and low-cycle regimes on the basis of pseudo-elastic stress range.

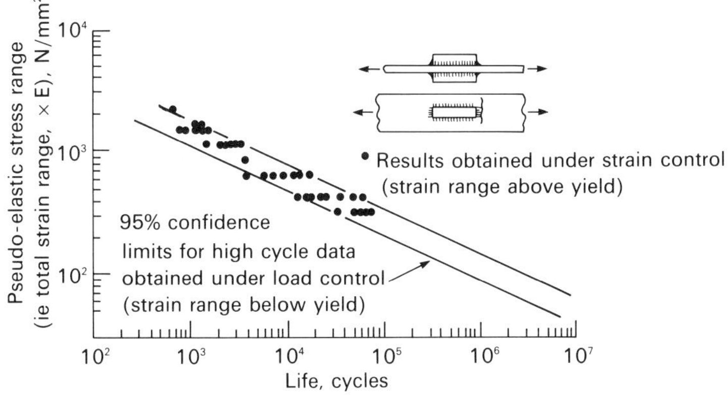

11 Correlation of low-cycle and high-cycle fatigue behaviour for fillet welds in steel.

Tubular connections

The increasing use of structural hollow sections, in vehicle chassis or lightweight space frames on the small scale and in offshore jacket structures on the large scale, has highlighted the potentially very severe stress concentration which arises at tubular intersections (nodes) where 'brace' members meet a continuous unperforated 'chord' member. Unlike the case of a nozzle in a vessel, the stress concentration is not due to an opening in the chord, but rather to the increase in stress which occurs in the chord as a result of local chord wall bending under loading from the brace. However, it is worth noting that external loading of a nozzle in a pressure vessel would produce the same effect in the vessel walls. Depending on the combination of brace and chord dimensions, local increases in stress compared with the nominal stress in the brace of up to thirty have been recorded. The variation in support between the brace and chord also gives rise to a stress concentration in the brace itself. Stress distributions in tubular joints have been widely studied and it has proved possible to derive a stress parameter, the 'hot spot stress', which takes into account the high local bending stress and the structural arrangement of the joint. Thus, this hot spot stress differs from the nominal stress normally used in conjunction with design S–N curves in that it includes the stress concentration due to joint geometry. As a result, the S–N curve for fatigue cracking of the chord of a tubular joint is relatively high, in fact coincident with that for transverse butt welds (Class D) for which the main source of stress concentration factor is simply the weld itself. Parametric equations for calculating the hot spot stress in a wide range of tubular connections are available in the published literature.

Misalignment

As discussed in Pt 1, misalignment of members at a joint, Fig. 12, particularly butt or cruciform joints, introduces a stress concentration if the joint is subjected to **axial** loading **transverse** to the line of misalignment. This stress concentration arises as a result of the superposition of local secondary bending stresses due to the misalignment on to the nominal stress field. The fatigue strengths of aligned and misaligned joints agree if the latter is expressed in terms of the resulting total stress range. Thus, the appropriate design S–N curve for the aligned joint can still be applied as long as the stress concentration due to misalignment is taken into consideration when calculating the stress range. This can be achieved using an appropriate stress magnification factor K_m but, since secondary bending stresses only arise under applied axial stresses, in the general design case it will be more convenient to calculate that secondary bending stress and add it to the applied stresses (which may be a combination of axial and bending stresses). Formulae are given in Table 3 for calculating the secondary bending stress S_b for a wide range of types of misalignment, in terms of the axial component of applied stress S_a. The equivalent stress magnification factor is $K_m = 1 + \frac{S_b}{S_a}$.

Note that these formulae make no allowance for restraint on the

Table 3 Formulae for calculating the secondary bending stress (at the weld toe unless stated otherwise) S_b due to misalignment in joints under transverse axial stress S_a. [Note that, if required, $K_m = 1+S_b/S_a$].

Type	Figure	Bending stress, S_b
Axial misalignment in butt or cruciform joints between flat plates:		
- of equal thickness	12(a) & (c)	$\frac{3e}{B} \cdot S_a$
- of unequal thickness	12(d)	$\frac{6e}{B_1} \left[\frac{B_1^{1.5}}{B_1^{1.5} + B_2^{1.5}}\right] S_a$ $B_1 < B_2$
Axial misalignment in fillet welded cruciform joints *	12(a)	$\frac{e}{B+H} \cdot S_w$ where S_w = stress on weld throat H = weld leg length
Axial misalignment in butt welded seam of pressurised vessel or pipe †	12(c) & (d)	$\frac{6e}{B_1(1-\nu^2)} \left[\frac{B_1^{n}}{B_1^{n} + B_2^{n}}\right] S_a$ where ν = Poisson's ratio $B_1 =< B_2$
Angular misalignment in butt or cruciform joints between flat plates. Boundary conditions equivalent to:	12(b)	
- fixed ends		$\frac{3}{4} \propto \frac{L}{B} \cdot S_a$
- pinned ends		$\frac{3}{2} \propto \frac{L}{B} \cdot S_a$
Angular misalignment in seam of pressurised vessel or pipe ‡		$\frac{6d}{B} \cdot S_a$

Key: * Relevant to fatigue failure in weld throat from root
† $n = 1.5$ for circumferential seams and seams in spheres
$n = 0.6$ for longitudinal seams
‡ d = deviation from true circular shape. Assumes pinned end boundary conditions

joint. In practice, the value of S_b will depend on the ability of the joint to rotate under the induced bending moment. This depends on loading and boundary conditions, section shape and the presence of other members which provide local stiffening. Special analysis is usually required to quantify these effects.

The need to consider the effect of misalignment at the design stage will depend on the type of misalignment, the extent to which an allowance is already made for misalignment in the design classification and on the workmanship standard to be used in the fabrication of the structure. Axial misalignment arising simply from a change in plate thickness at a butt joint is a design feature which should be considered when estimating the stress range to be used in conjunction with the design S–N curve. Table 3 includes relevant formulae. Similarly, axial misalignment will be an inherent feature of any lap joint. With regard to the design classification, there is a case for assuming that, when applied to transverse butt welds, the Class E design S–N curve already embodies an allowance for misalignment corresponding to $K_m = 1.3$, but this is not universally accepted in design rules. Thus, in general, the designer should consider the third point, the fabrication standard, and, at the very least, assume that misalignment up to the allowable level will be present and include its effect in calculations. If the allowance is then exceeded during fabrication, the misalignment may need to be re-assessed as a weld imperfection, as discussed in Chapter 4.

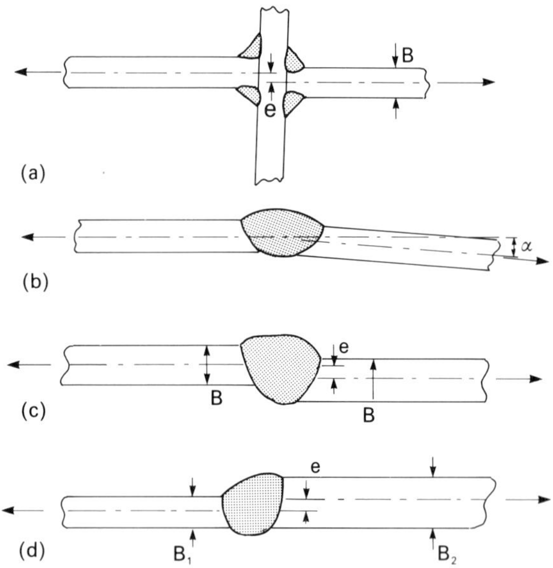

12 Examples of misalignment: (a) Axial, in cruciform joint; (b) Angular, in butt weld: (c) Axial, in butt weld; (d) Axial, stemming from thickness change at butt weld.

Chapter 3

Practical application of the rules

General design procedure

In general, structures should be designed to have a fatigue life which is equal to or greater than the anticipated service life. In some circumstances a decision may be taken to design for a fatigue life which is shorter than the service life, but this approach should be considered only if the structure, or the relevant part of it, can be inspected and repaired. It is also implied that the strength of the structure in the presence of any cracks will have been evaluated in advance.

Structures subjected to very high frequency loading, such as some rotating machinery or vibrating components, will probably have to be designed for infinite life. In other words, design stresses would need to be limited to S_0 for the relevant joint class.

The procedure, in summary, is then:

- Allocate the relevant joint classification to each joint in the component under consideration.
- Decide the acceptable probability of failure and hence the design S–N curve to be used. In most instances 2.3% probability of failure will be appropriate (M–2SD curve), but under non-critical conditions, or where inspection and maintenance are simple, the M–1SD curve (15.9% probability of failure) might be chosen. Conversely, in highly critical components or in situations where there is considerable doubt about the stress history, the M–3SD curve may be used.
- Establish the expected load or stress spectrum to which the component will be subjected during its life.
- Use the relevant design S–N curve to calculate either the total fatigue life of the component for a known stress spectrum or, more commonly, the maximum allowable stress range (or required cross-section of member) for a known load spectrum and required life.

Joints subjected to constant range loading

In practice, it is quite rare to find a structure or component subjected to

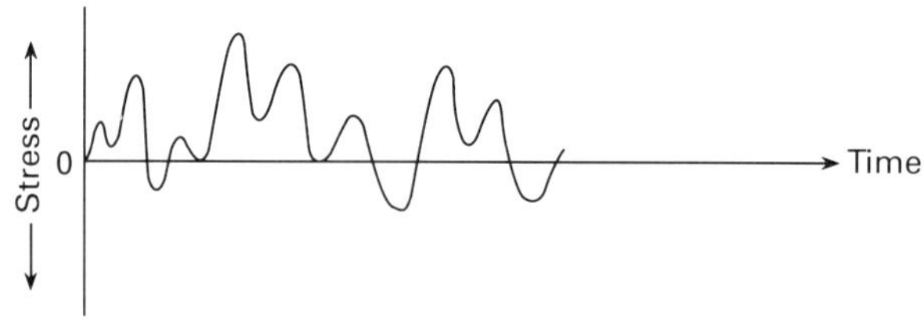

13 Variable stress range loading.

repeated loading of constant magnitude. However, when this occurs, fatigue design simply involves the direct application of the relevant design S–N curve. For the required fatigue endurance, the allowable working stress range should not exceed that defined using the equations in Table 2 for the relevant joint class and required probability of failure. For example, for a Class F detail which is required to withstand 10^6 repetitions of loading with a probability of failure of 2.3%, from the M–2SD Class F S–N curve, the allowable stress range is 86 N/mm^2. If the required life exceeds 10^7 cycles the requirement is for essentially infinite life and the allowable stress range cannot exceed S_0, that is 40 N/mm^2 on the Class F M–2SD curve. Conversely, if after preliminary static design a joint is found to be subjected to a cyclic stress range which is less than S_0, fatigue need not be considered further.

So far throughout the book the fatigue strengths of different welded joints have been discussed in terms of constant range loading. In most structures the usual service conditions give rise to variable loading giving stress ranges which vary, often randomly, Fig. 13.

The steps involved in fatigue design for variable range loading are described in the following section.

Random loading and cumulative damage

Service loading

Before any attempts can be made to obtain meaningful estimates of the fatigue behaviour of a structure or component, it is necessary to determine the stress spectrum which will be experienced in service. The importance of this task cannot be over-emphasised. As will be seen from the design S–N curves, fatigue life is proportional to stress range raised to the power of 3, or more for classes above D. Consequently, small errors in the estimation of S can lead to large errors in predicted life. The designer should aim to establish service loadings as accurately as possible, introducing conservative assumptions where doubts exist. In some situations, notably highway and railway bridges, standard loading rules exist which can be used to derive a stress specturm. It might even be possible to make reasonable assumptions as to the expected service history. For example, if the stress for a given load can be calculated and the load sequence follows a regular pattern, or a statistical distribution of loads can be assumed, the stress spectrum can obviously be deduced. However, the most reliable way to obtain a reasonable picture of the working stress variation is to take strain measurements, usually with electrical resistance strain gauges, on a prototype or a similar existing structure during the complete range to typical working operations.

Cumulative damage

The waveform produced under variable amplitude loading can be complex and, as seen earlier, bear little resemblance to that obtained under the constant amplitude loading conditions used to generate the fatigue test data upon which the design rules are based. The question arises: how can these data be used to design structures subjected to variable amplitude loading? It would clearly be ideal to test structures or components under simulated service loading conditions; indeed, since the advent of servohydraulic testing machines this may be feasible but, for practical and economic reasons, is rarely possible and instead the fatigue life under variable amplitude loading is usually calculated.

The first step is to break down the complex waveform into recognisable cycles, a process called 'cycle counting' as discussed below, so that the stress sequence may be described in terms of numbers of cycles, n_i, applied at different stress ranges, S_i (i = 1, 2, 3, etc.). The fatigue life of a joint subjected to this sequence is then estimated by summing the calculated fatigue damage arising from each block of cycles using a cumulative damage rule.

The most widely used rule and the one recommended for the design of welded joints is the Palmgren-Miner linear cumulative damage rule, commonly called Miner's rule. This states that the fatigue damage from any particular stress range is directly proportional to the number of cycles applied at that stress. For a particular stress range, S_1, the constant amplitude endurance, N_1, is a measure of the available damage (that required to cause failure) and therefore the fatigue damage as a result of S_1, applied for n_1 cycles, is n_1/N_1 of that needed to cause failure. Failure occurs under a sequence of different stresses when the sum of all the ratios n/N equals unity. That is

$$\frac{n_1}{N_1} + \frac{n_2}{N_2} + \frac{n_3}{N_3} + \ldots = \sum \frac{n_i}{N_i} = 1 \text{ at failure} \qquad [3]$$

Values of N_i are taken from the appropriate design S–N curve for each value of S_i, Fig. 14.

As an example of the use of Miner's rule, consider a structure which incorporates non-load-carrying fillet welded details (Class F). It is estimated that, during one year's service, the structure will be subjected to loads which give the following stress ranges in the region of the fillet welded details:

50 000 cycles of stress range 110 N/mm^2
120 000 cycles of stress range 80 N/mm^2
1000 000 cycles of stress range 50 N/mm^2

For 2.3% probability of failure, the values of N calculated using the equation for the Class F M–2SD curve are 4.75×10^5, 1.24×10^6, and 5×10^6 for the three stress ranges. Thus, in one year the fatigue damage accumulated is proportional to:

$$\sum\frac{n}{N} = \frac{5\times10^4}{4.75\times10^5}+\frac{1.2\times10^5}{1.24\times10^6}+\frac{10^6}{5\times10^6} = 0.4$$

so that failure would occur in 1/0.4 years, that is 2½ years.

It may be noted that N_i can be expressed in terms of S_i and the constants m and A given in Table 2

$$N_i = \frac{A}{S_i^{\,m}} \qquad [4]$$

so that Miner's rule could be stated as

$$\Sigma S_i^{\,m} n_i = A \text{ at failure, or} \qquad [5]$$

$$\Sigma S_i^{\,m} n_i \ngtr = A \text{ for survival}$$

Note that the fatigue damage due to a block of n_i cycles at S_i is proportional to $S_i^m n_i$. The application of Miner's rule is discussed further in Appendix 1.

Many investigations have been carried out to test the validity of Miner's rule and there is some diversity in the values of $\sum\frac{n}{N}$ (which should be unity at failure if the rule is correct) actually obtained. Until

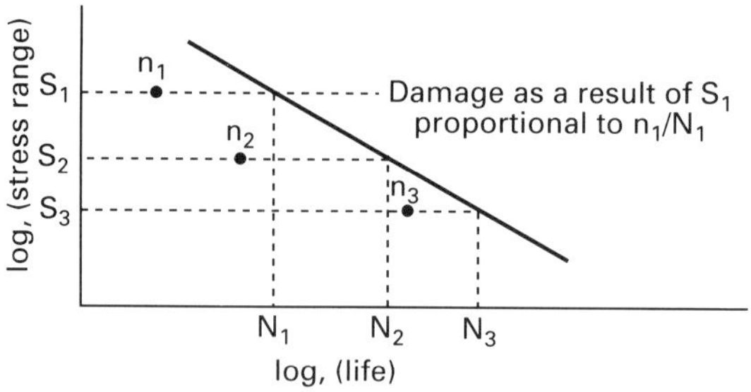

14 Application of Miner's rule.

relatively recently, most tests were performed under block programme loading (Fig. 15). This type of loading is a simplification of actual service loading, containing the same mix of stress ranges and the number of times they occur but applying them in blocks of constant stress range. The results of block programme load fatigue tests have indicated that, in general, Miner's rule is safe ($\sum \frac{n}{N} > 1$ at failure of the test specimen). Tests performed under more realistic spectrum loading conditions (e.g. simulating or even reproducing spectra of the type illustrated in Fig. 13) have also increased confidence in the use of Miner's rule. In particular, if the applied stress spectrum is one containing occasional relatively high tensile stresses ('overloads'), the fatigue life can greatly exceed that estimated using Miner's rule, because of the occurrence of 'crack growth retardation'. This phenomenon is thought to stem from the introduction of compressive residual stresses at the fatigue crack tip. Thus, under tensile stress, plastic straining occurs local to the crack tip; on release of the stress the surrounding material attempts to cover elastically and in so doing subjects the yield zone to compressive stress. The presence of this compressive residual stress then inhibits the damaging effect of subsequent (lower) applied stresses, Fig. 16. In other words, overloads producing tensile

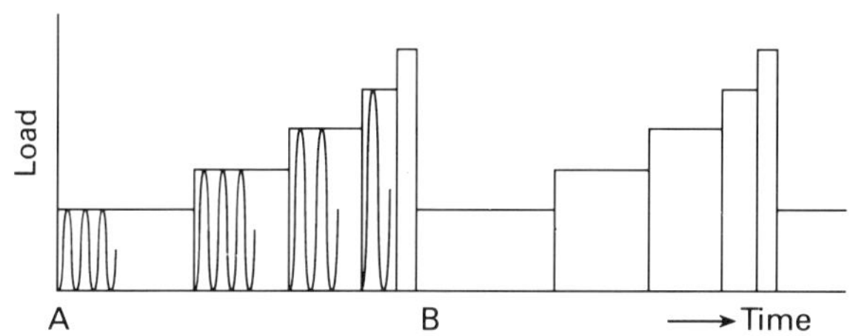

15 Example of a block programme load sequence. Sequence A to B is applied repeatedly until the test specimen fails.

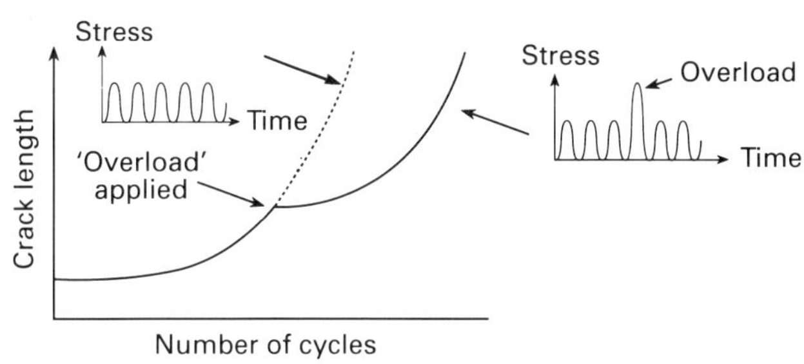

16 Schematic illustration of decrease in crack growth rate following application of tensile overload cycle.

stresses can be beneficial to fatigue life. A similar mechanism probably explains why actual lives exceed those estimated using Miner's rule in tests performed under block programme loading, the 'overload' effect occurring when there is a step change from one stress level to the next at the end of a block, Fig. 15. In contrast, Miner's rule proves to be reasonably accurate if the applied stress spectrum resembles narrow-band random loading, as might arise as a result of resonance-induced vibration of the structure. The significant characteristic of such a spectrum is that the stress range changes gradually, with the result 'crack growth retardation' is not significant.

From the practical viewpoint fatigue test results obtained under block programme loading must be treated with caution, since the actual service stress spectra which are being simulated may not actually contain the step changes included in the test which can give rise to crack growth retardation. Indeed, in general there is only limited scope for taking account of the benefit of the crack growth retardation effect seen in any variable stress range fatigue tests, unless tensile overloads are deliberately applied in service, because few service stress spectra are sufficiently well defined to be sure that 'overloads' will actually arise.

Unsafe estimates of fatigue life using Miner's rule have also been observed for some stress spectra. These include stress cycles which contain small fluctuations, such as might occur as a result of vibrations being superimposed on to a lower frequency stress fluctuation, Fig. 17, and under some random and pseudo-random load sequences. The reasons for this are not fully understood and further research is still needed in this important area. However, some progress has been made by Gurney (see Bibliography) in deriving an empirical relationship which can be more accurate than Miner's rule, particularly for loading conditions under which Miner's rule is unsafe.

For an applied stress spectrum expressed in terms of proportions p_i of the maximum stress range S_{max} and broken down into repeatedly

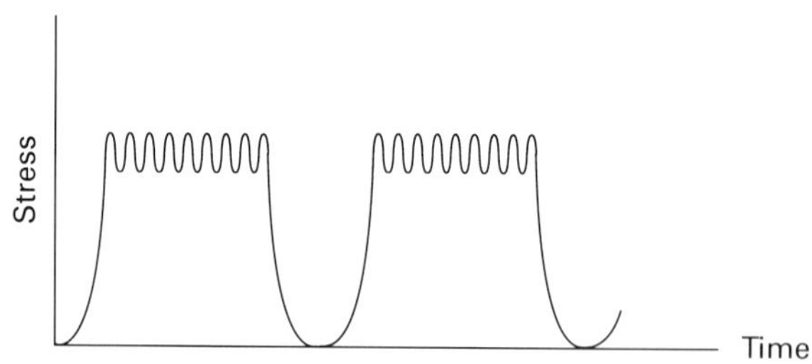

17 Stress spectrum containing small fluctuations superimposed on to major stress cycles.

applied blocks of cycles with each block containing only one application of S_{max}, Gurney's rule states:

$$N_B = \prod_{i}^{n} \left| \frac{N_{E_{i-1}}}{N_E} \right|^{p_i} \cdot N_C \qquad [6]$$

where $N_B =$ predicted life in blocks

$N_C =$ constant amplitude life under S_{max}

$N_{E_i} =$ number of cycles per block at stresses $\geqslant p_i S_{max}$

This can also be expressed (using natural logarithms, ln, i.e. $\log_e$):

$$\ln\left(\frac{N_B}{N_C}\right) = -\text{ (area under the } p_i \text{ versus } \ln N_{E_i} \text{ exceedance diagram)}$$

If the repeatedly applied block of cycles contains more than one cycle at S_{max} or the spectrum contains no pattern of repeatedly applied blocks and, again, more than one cycle at S_{max}, all but one of these cycles would be assumed to adopt a value $S = 0.99\ S_{max}$.

Miner's rule is still the only cumulative damage method included in design rules, although recognition of the fact that in some circumstances it could be unsafe has in some cases prompted the adoption of a summation value of less than unity for added safety. Future research may show that there is scope for an alternative cumulative damage method, perhaps composed of a combination of Miner's rule and Gurney's rule.

Miner's rule can also be unsafe if the loading conditions are such that a majority of applied stress ranges are near the relevant value of S_0. If they are below this, even just below, they are disregarded in a Miner's rule calculation since

$$\frac{n}{N} = \frac{n}{\infty} = 0$$

However, in reality, when crack propagation occurs under the action of higher stress ranges in the spectrum, such stress ranges can contribute to propagation and therefore reduce the fatigue life. Modified calculations may be made to account for this effect based on the extrapolation of the design S–N curve from S_0 at the same slope to a lower fatigue limit S_0 or at a different slope, as discussed in Appendix I.

Cycle counting

To apply Miner's rule, service stress conditions expressed in terms of numbers of complete cycles at particular stress levels are required, essentially in the form used in block programme fatigue testing in, Fig. 15. In practice, situations may arise when the service loading conditions are readily reduced to such a form, as indicated for the tension flange of a crane in Fig. 18. However service load conditions are usually more complex than this, Fig. 19. This time the spectrum does not contain well-defined complete cycles and, to apply Miner's rule, it is necessary to restate the complex spectrum in simpler terms, the process of cycle counting.

Several cycle counting techniques are available but little is known about their relative merits. The two which are probably most relevant to loading applied to as-welded joints, where stress ranges are of interest, are the range-pair and rainflow (or reservoir) methods, the latter being recommended in BS 5400. Their application is described in Appendix II and Fig. 20 shows the results of the two methods for a sample

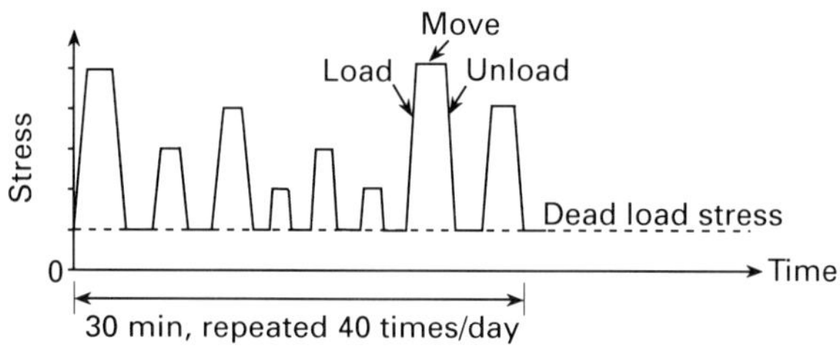

18 Stress spectrum for tension flange of crane which moves four known weights across workshop.

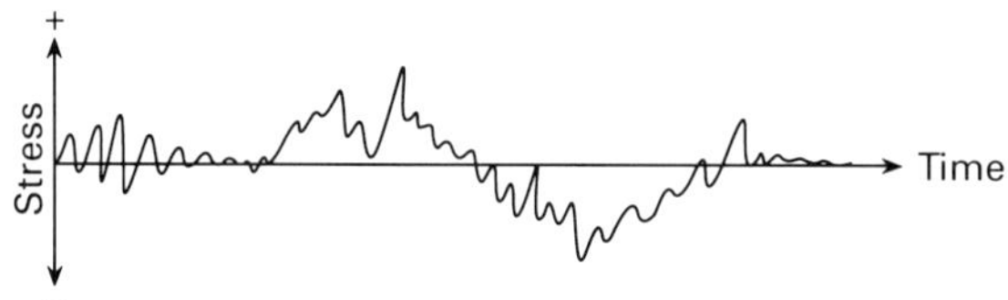

19 Stress spectrum for mobile crane.

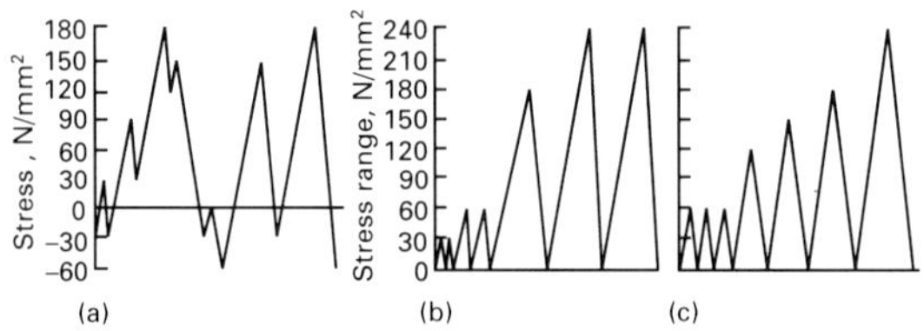

20 Comparison of results of rainflow and range-pair cycle counting methods: (a) Spectrum, equivalent to sequence according to; (b) Rainflow, and; (c) Range-pair counts.

waveform. As will be seen, different methods applied to the same waveform can lead to supposedly equivalent blocks of constant stress range which actually would produce quite different amounts of fatigue damage. The relative fatigue damage predicted by the two methods can be compared by calculating $\Sigma S^m n$, as noted earlier. For example, assuming that a detail in Class D or lower is considered, m=3 and the range-pair method predicts damage proportional to 2.54×10^7, while the rainflow method predicts damage proportional to 3.4×10^7, i.e. 34% more. It is not possible to make recommendations about the selection of a cycle counting method at present. A conservative approach would be to try several methods and choose the one which resulted in the greatest fatigue damage.

Load and stress distributions

The cycle counting procedure yields data in the form of load or stress range against number of cycles occurring at that range. This can be represented as a histogram. The original spectrum sample should be produced over a sufficiently long period to ensure that the histogram gives a representative distribution of loading for the structure in question. For cumulative damage calculations, distributions of loading have to be converted to distributions of stresses.

The histogram, Fig. 21, can be presented in a more general form by plotting

$$\frac{\text{frequency of occurrence}}{\text{stress range interval}},$$

called the probability density, against stress range to give the curve shown. If an equation can be fitted to the curve it could be used when carrying out cumulative damage calculations by computer. In practice

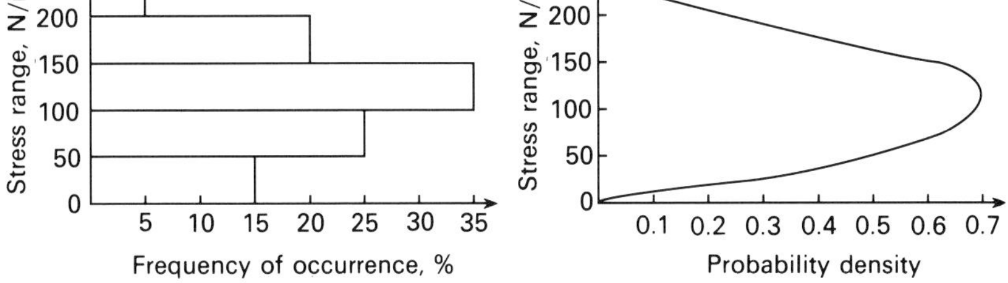

21 Stress spectrum.

it is often possible to represent random load spectra by general mathematical expressions, such as Rayleigh probability curves.

Design of transverse load-carrying fillet welds

As discussed in Pt 1, joints containing load-carrying fillet welds can develop fatigue cracks which propagate in the:

- Plate from the weld toe, or
- Weld from the weld root.

The fatigue strength obtained for weld failure can be much lower than that obtained for plate failure and the optimum situation is one in which the strengths have been made equal, together with an equal probability of failure by either mode. The experimental basis of such design, for some fillet welds, is shown in Fig. 22.

One approach would be to use the design S–N curve for plate failure (Class F_2) and ensure that the optimum geometry was achieved by reference to curves of the kind given in Fig. 22. Alternatively, a simplified

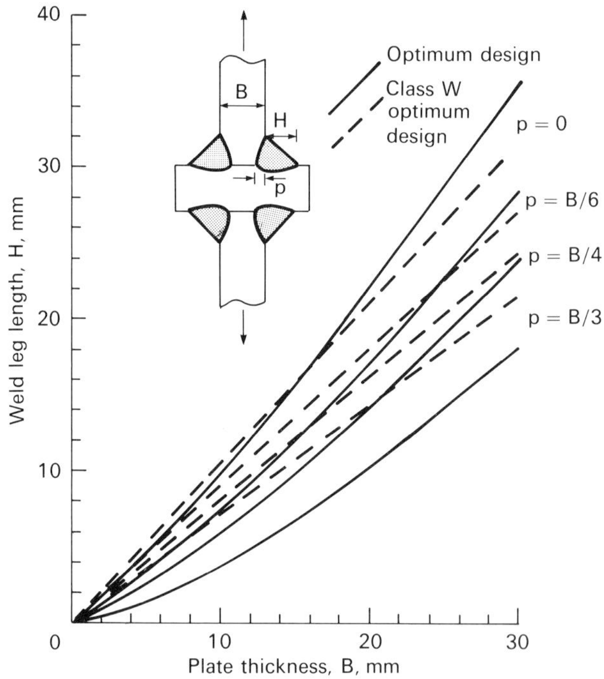

22 Conditions of optimum geometry of transverse load-carrying fillet welds.

approach would be to confine the design rules to fillet welds and introduce the correct weld size : plate thickness ratio by using an S–N curve based on the stress in the weld. This is the approach used in the present rules, the classification being W, based on the stress in the weld throat. Class W covers both transverse and longitudinal welds and the design stresses have been selected on the basis of test data for both types of joint. In transverse welds the design stresses for Classes F_2 and W are such that the fatigue life is equal if the weld leg length, H, is approximately the same as the plate thickness, B, which is reasonably compatible with the optimum design, as seen in Fig. 22 ($P = 0$).

A consideration in the use of the Class W design stresses for transverse welds is that they do not necessarily allow advantage to be gained from the use of partial penetration welds. The optimum weld size : plate thickness relationship for three depths of weld penetration is shown in Fig. 22, together with the (dashed) curves derived on the basis that Class W and Class F_2 give the same fatigue lives, assuming that the weld throat size = $(H+p)/\sqrt{2}$. As depth of penetration increases, so the difference between curves increases. Thus, for a weld with one-third plate thickness penetration between 19 mm thick plates, the required weld leg length to ensure that the Class W life is not less than the Class F_2 life is 14 mm. However, it will be noted that, according to the relevant optimum curve, based on test results, a weld size of 9.5 mm would suffice.

Stress-relieved joints

A major difference between the present rules and fatigue design rules for unwelded components is that the applied stress ratio is disregarded and design is based simply on the total applied stress range. This approach is used because it must be assumed that as-welded joints will contain tensile residual stresses of yield strength magnitude. This will not be so with stress-relieved joints, although it cannot be assumed that residual stresses will be relieved completely. Nevertheless, the reduction in residual stress level will affect fatigue strength if the loading gives all or partly compressive stresses. Further research is required before detailed design rules for stress-relieved joints can be drawn up. Even so, the practical application of such rules could well be seriously limited by the problem of ensuring that residual stresses have really been relaxed as a result of stress relief treatment. Very high tensile residual stresses, approaching yield, have been measured in large stress-relieved structures.

It may be noted that the task of analysing the stress spectrum for a given structure is more simple when it involves the identification of stress ranges only; it could well be considered that the kind of increase in allowable stress range which might prove to be justifiable for the stress-relieved joint would not justify the extra complexity of having to determine stress ratio in addition, particularly if the applied range is predominantly tensile.

High strength and special steels

It was stated in Pt 1 that the fatigue strengths of welded details, characterised by mean S–N curves, were common to structural steels and independent of their tensile properties. It follows that the design curves are also common. In fact, this situation applies to welded steels in general, including high strength or special steels which can otherwise be welded satisfactorily, Fig. 23.

Such behaviour was explained in Pt 1 by reference to the initiation and propagation of fatigue cracks. It is observed that the tensile strength of steel has little effect on the **rate of propagation** of a crack. The fact that unwelded specimens show benefit from increases of tensile strength must therefore derive from the existence of crack initiation, or incubation, periods in addition to that required for propagation; equally, the constant fatigue strength of welded details implies that their cyclic life is determined essentially by propagation alone. This has been observed to be the situation for joints in Class D and below. Such behaviour is attributable to the presence of pre-existing cracklike

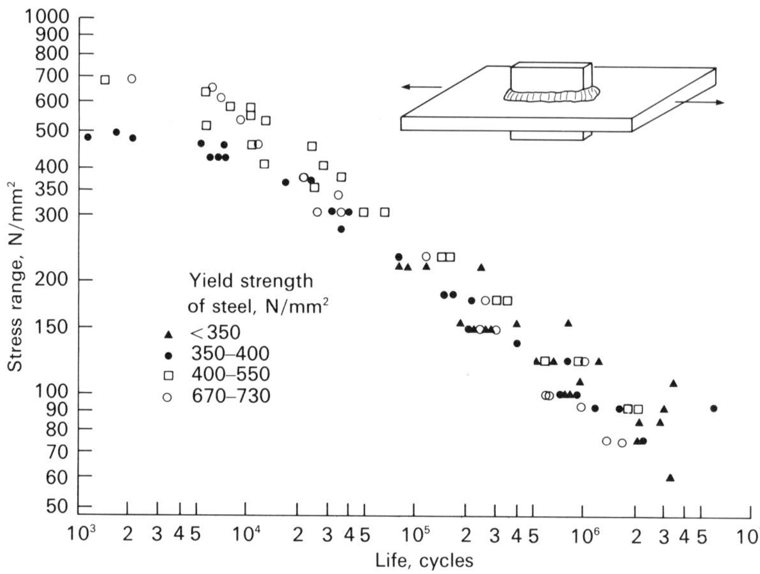

23 Fatigue test results from fillet welds in various strengths of steel.

flaws, such as the inherent intrusions at weld toes, which, in effect, represent a condition in which crack initiation has been achieved.

Welded high strength steels offer, therefore, no intrinsic advantage in terms of fatigue strength, and it is necessary to assess design criteria carefully to determine when they may be used with benefit in the context of fatigue loading. The considerations can be outlined by referring to an S–N design curve for a given class of joint, Fig. 24, as might be relevant to the member or structure concerned. For simplification, the load cycle is taken to be repeated tension, or R=0, so that stress range and S_{max} are coincident in value. This permits stresses S_1 and S_2 to be entered, representing the maximum permissible **static** tensile stress for, say, an ordinary grade of mild steel, S_1, and a higher strength steel, S_2. Such stresses might typically be of the order of

$$\frac{\text{yield strength}}{1.7}$$

each value representing the overriding maximum stress for the corresponding grade of steel.

Stresses S_1 and S_2 therefore set limits to the permissible peak stress, S_{max}, of an applied fatigue cycle and it is evident that this can affect the selection of steel in certain circumstances. If, for example, the design objective involves the requirement of limited endurance, N_2, it is clear that the substitution of the higher strength steel enables the fatigue design stress to be increased from $S_{max}=S_1$ to $S_{max}=S_2$. Conversely, this benefit would be eliminated if a life requirement of N_1 cycles was specified, when $S_{max}=S_1$ would apply, irrespective of which of the two steels was used. In this respect it will be noted that the ability to employ an increased stress range with the high strength grade is extended if the S–N curve is shifted to the right, i.e. by the adoption

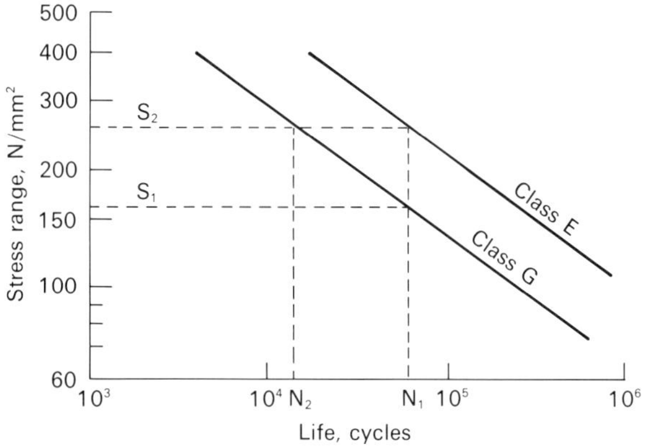

24 Design S–N curves illustrating utilisation of high strength steels.

of the highest feasible class of joint. For example, the high strength steel could be used to advantage even for the higher life requirement, N_1, if the limiting weld details were Class E.

The same considerations apply when loading conforms with other stress ratios, although increasingly as R diminishes the allowable fatigue stress range will predominate rather than observance of the upper limit cut-off stress. Where the mean stress is high compared with the cyclic range, or where in spectrum loading it can be foreseen that discrete high loads will need to be accommodated, the question of providing what might be regarded as the basic strength of higher tensile steel, working at levels above the S_1 limit for mild steel, will, however, be reintroduced, in combination with the relevant limits imposed by the fatigue design curves.

Throughout this discussion the use of carbon-manganese or low alloy ferritic steels has been understood. It is the case that the design rules have been built up substantially from evidence of the fatigue behaviour of steels of this kind, but there are grounds for making the assumption that others, such as the austenitic stainless steels, would show similar behaviour and in-air strength when welded.

Welded joints in aluminium alloys

A general description of the fatigue characteristics of welded joints in steel, as given in Pt 1, is equally applicable to joints in weldable aluminium alloys. Stress concentration effects and crack locations are similar and fatigue strengths are again independent of the tensile properties of the material. Actual strengths are, of course, different from those quoted for steel. Stresses required to produce the same rate of fatigue crack propagation in steels and aluminium alloys are roughly in proportion to the respective elastic moduli, i.e. three to one (see Fig. AIII.4), and therefore the fatigue strengths of welded joints, or the design S–N curves for joints, can be taken to have the same relationship, Fig. 25.

The use of weldable aluminium alloys in fatigue-loaded structures is considered in detail in BS 8118.

Effect of environment

The fatigue design rules are based on test data obtained in air at ambient temperature and therefore do not allow for any adverse influence of environment. In practice, welded structures may be required to operate in all manner of environments – corrosive, elevated or sub-zero temperatures – and each one may have a significant influence on allowable fatigue stresses.

Corrosion fatigue

An environment may produce negligible general corrosion of a structure but the presence of crevices or cracks can lead to intensive localised attack. In some circumstances a corrosive environment exists only because there is a crevice in which moisture can be trapped. It will be appreciated that cracks initiated by fatigue loading will provide ideal sites for accelerated corrosive attack. When fatigue loading and corrosion are combined the resulting problem can be more severe than when the two effects are separate.

The designer should be aware of two possible effects of a corrosive environment which could mean that the design rules presented earlier are not applicable. Firstly, when the fluctuating stresses in a structure are high enough to allow fatigue cracks to propagate from stress concentrations, corrosive reactions may be set up within the cracks causing more rapid propagation. Thus the corrosion fatigue strength of a welded joint can be less than that in air. Secondly, although stresses below the fatigue limit would not be expected to lead to fatigue failure in air, localised corrosive attack at stress concentrations and surface defects, such as the weld toe area, can increase the severity of the stress concentration and enable fatigue cracking to occur. Thus, the constant amplitude fatigue limit is reduced. Similarly, the development of corrosion pits in dressed surfaces can virtually eliminate the

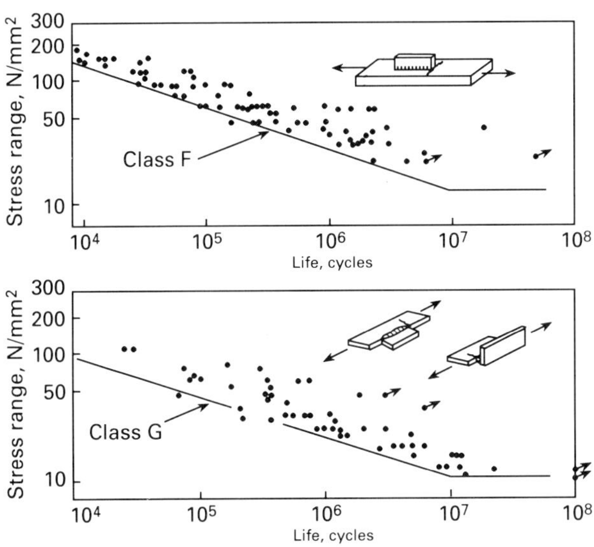

25 Fatigue data for welded aluminium alloy (0) compared with relevant design curve (M–2SD) for steels plotted in terms of stress range/3.

improvement in fatigue strength which the dressing of welds otherwise confers under in-air conditions.

The corrosion fatigue strength of welded joints depends upon many factors and it is not possible to produce general design rules. When considering test results to obtain guidance about the possible reduction in fatigue strength of a joint in a particular environment, the designer should be aware that the environment, temperature, frequency of cyclic loading, and waveform shape can all be significant and should reproduce as closely as possible the service conditions of the joint.

The importance of frequency can produce particular problems. Corrosion depends upon time which, unlike fatigue, cannot be accelerated, so that some tests may need to occupy several years before useful results are obtained.

A specific case which has been investigated quite extensively is the North Sea environment experienced by offshore structures, and some design recommendations have been made. Strictly speaking, these are only applicable to structural carbon-manganese steels in seawater temperatures between about 5 and 10°C and subjected to cyclic loading corresponding to the average wave frequency of 6 cycles per minute. For freely corroding welded joints the fatigue life is reduced by a factor of around 2 and the endurance limit is so depressed that, for design purposes, it is ignored. The beneficial effect of cathodic protection measures, to reduce corrosion, is not clear. Fatigue tests on simple welded joints indicate that in-air performance is restored. Tests on more realistic tubular joint specimens, in which the loading conditions are more complex, are less conclusive. This situation arises partly because the corrosion fatigue behaviour of the real structure has not been reproduced in the simple welded joint, the latter giving more favourable results. This illustrates the importance of two other factors which should be reproduced as closely as possible in a corrosion fatigue test, namely the geometry and loading.

As with other aspects of corrosion and fatigue, attention to detail design in the elimination of notches, discontinuities, and crevices where possible is essential.

Elevated temperature fatigue

As in corrosion fatigue it is not possible to lay down hard and fast rules about the effect of elevated temperature on the fatigue strengths of welded joints. Depending upon the material, temperature, and environment, fatigue strength may be more or less than that at ambient temperature. Unlike fatigue at ambient temperature, cyclic frequency and waveform shape can also influence fatigue behaviour. Corrosion fatigue strength tends to decrease with an increase in temperature. At temperatures in the creep regime for the particular material considered, fatigue may be insignificant compared with damage arising from creep crack growth.

Guidance is given in some codes. In general, it is assumed that the same fatigue design rules apply for structural carbon-manganese steels operating at temperatures up to 375°C, for austenitic stainless steels operating at temperatures up to 430°C and for aluminium alloys operating temperatures up to 100°C. However, in other cases, including at temperatures within the ranges quoted, if the environment is also corrosive, specialist advice should be sought.

Low temperature fatigue

Although comparatively little testing has been carried out, it seems clear that a reduction in temperature to below ambient generally reduces the rate of propagation of a fatigue crack in steel and hence increases fatigue life. However, the fatigue strength of a joint may still be reduced if the fracture toughness of the material is such that brittle fracture occurs from a small fatigue crack early on in the fatigue life. As a general guide, the critical crack size would need to be less than half the thickness of the section through which the fatigue crack was propagating for the fatigue life to be reduced significantly. However, even if the resulting fatigue life was adequate, the catastrophic nature of a final failure by brittle fracture may be unacceptable.

Chapter 4

Significance of weld imperfections

Virtually every welded joint contains stress concentration features other than welding flaws and therefore the incidence of such flaws may or may not be of consequence depending upon their severity. In practice, it may be noted that the majority of cases of fatigue cracking in service are associated with the inherent stress concentration features of the basic welded joint geometry, and only a relatively small number with welding flaws.

Many codes quote acceptance levels for welding imperfections but usually these are related to quality control and are therefore arbitrary and generally conservative. They will frequently need to be taken into account, but in some circumstances an ability to determine on a more rational basis whether or not a flaw should be rejected or accepted offers substantial advantages. A rational basis for the assessment of imperfections is 'engineering critical assessment' (ECA) based on fitness for purpose, whereby a flaw could be accepted provided it could be shown that its presence would not reduce the fatigue strength of the welded joint below that required. The avoidance of unnecessary repair of welding flaws on the basis of ECA, and consequent savings in fabrication costs and delays to projects, has saved millions of pounds, notably in the offshore and power generation industries. The ECA procedures used in BS PD 6493 are briefly reviewed here.

ECA procedures for weld imperfections

The scheme referred to applies to imperfections in fusion welded joints, in material at least 10 mm thick, and is basically applicable to structural ferritic steels, austenitic stainless steels and aluminium alloys.

Welding flaws are classified as either planar (cracks, incomplete penetration, lack of fusion) or non-planar (slag inclusions, porosity). A third category is shape imperfections, which covers weld toe undercut and misalignment of a joint. The basis of the recommendations is different for the various types of imperfection. However, the fatigue strength of joints containing any of these types of imperfection can be expressed in the same way, using a grid of S–N curves relating to a range of quality categories.

Quality categories

The quality category S–N curves are shown in Fig. 26, the diagram having two stress range scales, one for steels and the other for aluminium alloys.

The quality category required in a particular instance is identified in the most simple procedure by locating the point on the diagram which corresponds to the stress range and life already determined in the design of the joint or joints to be incorporated. If this point lies between two curves, the required fatigue strength is then defined by the S–N curve above the point. Thus, if a steel joint is required to withstand 4×10^6 cycles of a stress range of 60 N/mm², the required quality category is Q2 and the S–N curve between Q1 and Q2 is relevant.

In practice, the fatigue strength required for the flawed joint being considered may be limited by adjacent details subjected to the same stress conditions or indeed by other stress concentrations in the same joint. This situation is illustrated in Fig. 27 which shows a steel girder detail assumed to be subjected to uniform bending moment, so that the welds in the flange can be taken to be in the same stress field. Clearly the fatigue strength of flaws in the splice butt weld need be no higher than that of the nearby fillet welds, which here fall into Class F. In the absence of the fillet welds the limiting case may be the butt weld toes (Classes E or D). To facilitate comparisons between the quality

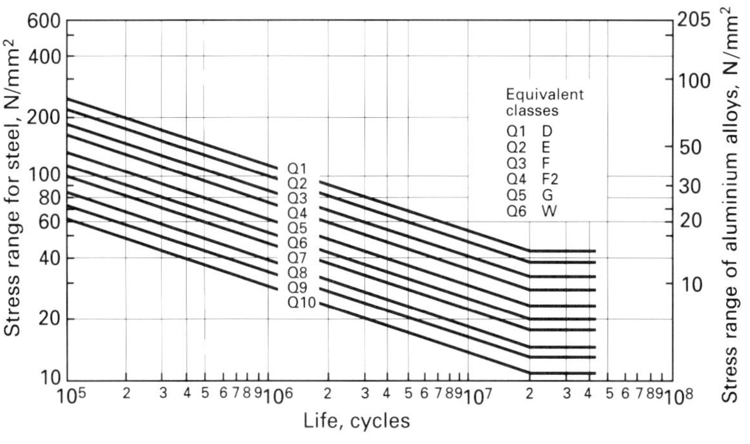

26 Weld quality categories.

categories and the design S–N curves, the S–N curves for quality categories Q1 to Q6 coincide with the Class D to W design curves (M–2SD) respectively. Thus a fatigue strength requirement corresponding to Class F, for example, is equivalent to Q3. Categories below Q6 are included because, of course, in some cases the required fatigue strength may be below that corresponding to the lowest design class.

Details of the quality category S–N curves shown in Fig. 26 are given in Table 4, including values of S_0, the modified endurance limit used in PD 6493 for cumulative damage calculations (see Appendix I).

For joints subjected to variable amplitude loading the required quality category is found by using the Palmgren-Miner cumulative damage rule. Recalling [5] and putting m = 3

$$\Sigma S_i^3 n_i = A' \text{ at failure} \qquad [7]$$

where A′ is the constant in the equation of the required constant amplitude S–N curve. Thus, the required quality category is that corresponding to the value of A in Table 4 which is closest to A′, with the condition $A \geqslant A'$.

An alternative method would be to substitute a co-ordinate on the constant amplitude S–N curve corresponding to A′: for example the stress range, S, which gives a life of 10^5 cycles. Then

$$\Sigma S_i^3 n_i = S^3 . 10^5$$

$$\text{i.e. } S = \left\{ \frac{\Sigma S_i^{\ 3} n_i}{10^5} \right\}^{1/3} \qquad [8]$$

From Table 4 the required quality category is that one having a stress range at 10^5 cycles next above the calculated value of S.

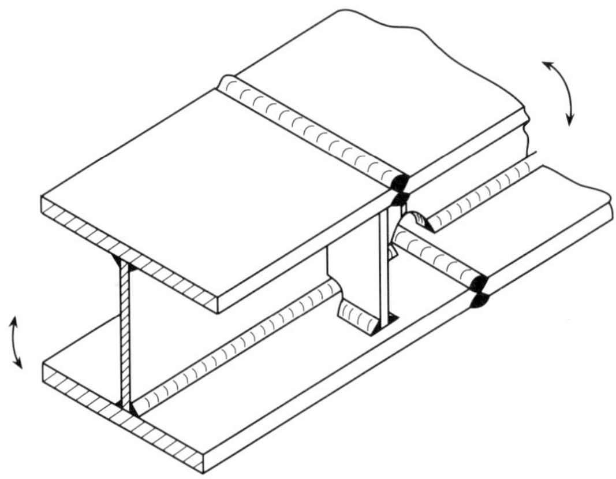

27 Steel girder with splice butt weld.

Table 4 Details of quality category S–N curves for as-welded joints in steel*

	Constants in equation $S^mN = A$		Stress range, N/mm²			
Quality category	m	A	10^5 cycles	10^6 cycles	10^7 cycles	S'_0, N/mm²
Q1	3	1.52×10^{12}	248	115	53	42
Q2	3	1.04×10^{12}	218	101	47	37
Q3	3	6.32×10^{11}	185	86	40	32
Q4	3	4.31×10^{11}	163	76	35	28
Q5	3	2.48×10^{11}	135	63	29	23
Q6	3	1.52×10^{11}	115	53	25	20
Q7	3	1.00×10^{11}	100	46	22	17
Q8	3	6.14×10^{10}	85	39	18	15
Q9	3	3.89×10^{10}	73	34	16	12.5
Q10	3	2.38×10^{10}	62	29	13	11

* For aluminium alloys divide A by 27 and stress range by 3 and assume $S'_0 = 0$

Shape imperfections

Misalignment

The welded joints most susceptible to misalignment are butt and cruciform joints, Fig. 12. Reference has already been made to the effect of misalignment as a source of stress concentration in such joints and the fact that fatigue test data obtained from aligned and misaligned welded joints are found to correlate in terms of the sum of the applied axial stress and the induced secondary bending stress (Pt 1). In view of this, the total stress can be used directly in conjunction with the appropriate design S–N curve for the aligned joint to assess the effect of misalignment. Table 3 includes formulae for calculating the stress magnification factor K_m in a variety of types of misaligned joints. Alternatively, in an ECA of misalignment, allowable values of K_m resulting from all types of misalignment present in the joint, to achieve fatigue strengths corresponding to the various quality categories, are given in Table 5. No misalignment (i.e. $K_m = 1$) is allowed for the quality category which corresponds to the normal design class for the joint and the allowances for lower quality categories are simply the factors by which the stress ranges corresponding to those categories must be multiplied to give the design class quality category. For example, in a Class E butt weld, $K_m = 1$ for Q2, for which S = 218 N/mm², then

$K_m = \frac{218}{185} = 1.18$ for Q3, $\frac{218}{163} = 1.34$ for Q4, etc. In the case of a cruciform fillet welded joint, the effect of misalignment must be assessed with respect to possible toe failure, based on K_m at the weld toe in the plate,

Table 5 Acceptance levels for misalignment

Quality category of misaligned joint	Allowable K_m					
	Design class	D	E	F	F_2	W
	Detail			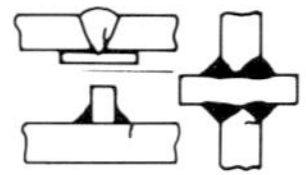	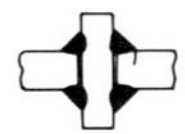	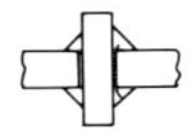
Q1		1.0	—	—	—	—
Q2		1.14	1.0	—	—	—
Q3		1.34	1.18	1.0	—	—
Q4		1.52	1.34	1.13	1.0	—
Q5		1.84	1.61	1.37	1.21	—
Q6		2.16	1.88	1.61	1.42	1.0
Q7		2.48	2.18	1.85	1.63	1.15
Q8		2.92	2.56	2.18	1.92	1.35
Q9		3.40	2.99	2.53	2.23	1.58
Q10		4.00	3.52	2.98	2.63	1.85

Note: (a) K_m = magnification factor due to any type of misalignment including combinations of more than one type.
(b) Assessment of class W joints refers to possible fatigue cracking in weld and is based on stress range on weld throat.

and possible weld throat failure, based on K_m at the weld root related to the stress range on the weld throat (see Table 3).

As misalignment only gives rise to secondary bending stresses in joints subjected to axial loading, joints subjected to bending are not sensitive to the imperfection. Similarly, in joints subjected to a combination of axial and bending stresses, K_m applies only to the axial stress component.

Clearly, misalignment may be present in any joint which is being assessed for other types of flaw. In such cases, the stress range used in the assessment should include any secondary bending stress due to misalignment, assuming that this bending stress varies linearly through the section thickness and is zero at the neutral axis. Thus, it will be particularly important to take account of misalignment in assessments of surface flaws.

Undercut

Fatigue test data for fillet and butt welds with known levels of undercut have been used to deduce the acceptance levels corresponding to each quality category given in Table 6. Rather few test data are available and therefore the recommendations are deliberately conservative, with the requirement that undercut deeper than 1 mm must be assessed as a planar flaw.

Longitudinally loaded joints

Since misalignment and undercut are only important from a fitness-for-purpose point of view in joints loaded in the transverse direction, no limits to the extent of these imperfections are given for continuous welds stressed in the longitudinal direction.

Table 6 Acceptance levels for weld toe undercut

Quality category	Depth of undercut / Plate thickness	
	Butt welds	Fillet welds
Q1	0.025	—
Q2	0.05	—
Q3	0.075	0.025
Q4	0.10	0.05
Q5	0.10	0.075
Q6–Q10	0.10	0.10

Note: Maximum depth of undercut in any thickness = 1mm.

Non-planar flaws

Fatigue test data for butt welds containing embedded porosity or slag inclusions have been interpreted on the basis of the set of quality category S–N curves. Hence the tolerable flaw levels to achieve particular quality categories have been identified, as summarised in Table 7.

It will be observed that limits for slag inclusions are given in terms of length. This refers essentially to the length of an inclusion measured in a direction normal to the applied principal stress, where necessary taking account of interaction with adjacent inclusions, as detailed in Fig. 28. The beneficial effect of postweld heat treatment on the slag inclusion limits derives from the release of hydrogen and not to the relief of residual stresses.

With regard to porosity, the radiograph reference area is that bounded by the maximum width of the weld and the length of weld affected by the porosity. Apart from strength considerations, the limits specified diminish the possibility of the porosity masking more harmful defects.

The limits given in Table 7 apply to butt or fillet welded joints in both steels and aluminium alloys, noting of course that the allowable stresses for aluminium alloys are lower, Fig. 26 and Table 4. Surface-breaking flaws are excluded: they should be treated as planar flaws. Similarly, multiple inclusions may have to be regarded as planar flaws, Fig. 28, and if there is any doubt about the type of an embedded flaw, it should be treated as planar.

Finally, it would be necessary to assess any flaw as planar if it was required to determine the progress of the fatigue crack which propagated from it. Such a situation would arise if the growing fatigue crack could initiate brittle fracture, or it was required to establish inspection

Table 7 Limits for non-planar flaws in steel and aluminium alloy joints

	Allowable length of slag inclusion,* mm		Limits of porosity	
Quality category	As-welded	Postweld heat treated	% of area on radiograph	Individual pore diameter
Q1	2.5	19	3	6mm or ¼ plate thickness if less
Q2	4	58	3	
Q3	10	Continuous	5	
Q4	35	Continuous	5	
Q5	Continuous		5	
Q6–Q10	Continuous		5	

* Tungsten inclusions in aluminium alloy welds need not be considered: their presence does not affect fatigue strength

periods. The acceptance limits are based on fatigue lives which effectively correspond to the attainment of through-section fatigue cracking and therefore they are not directly applicable for other failure criteria or for judging the fatigue crack size at various stages during the fatigue life.

Planar flaws

The treatment of embedded non-planar flaws was simplified by the fact that flaw severity could be characterised using one dimension, i.e. inclusion length and projected area of gas pores, with the result that their fatigue strengths could be readily represented by a few S–N curves based on test data. However the treatment of planar flaws is more complex, since the severity of the flaw depends not only upon its size but also its shape and location, all of which may vary from one weld to the next. These factors can be taken into account if the fatigue strength of the flaw is calculated using fracture mechanics (see Appendix III).

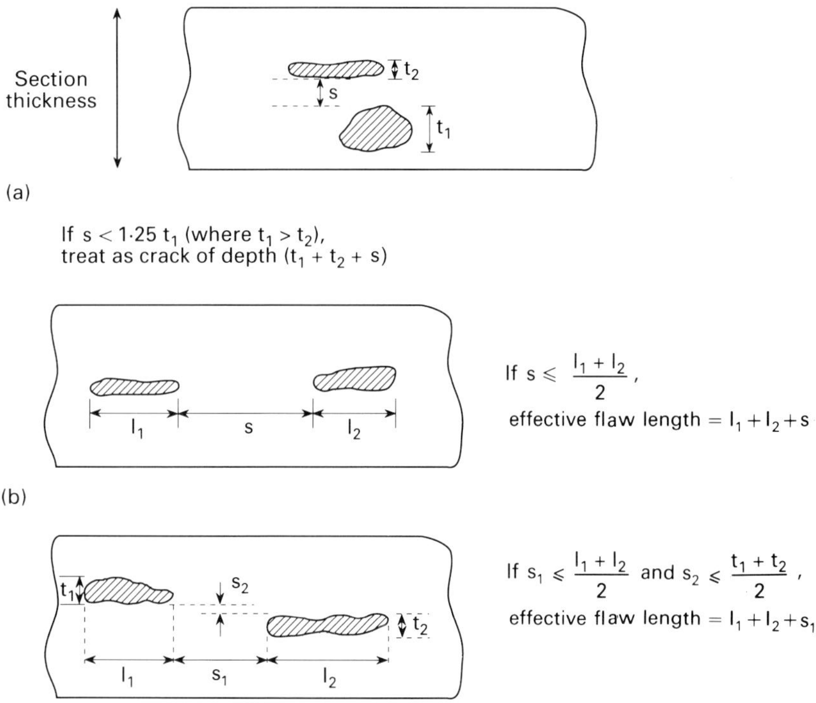

28 Interaction of slag inclusions: (a) Coplanar embedded slag inclusions; (b) Effective length of multiple slag inclusions in same or different plane(s).

The basis of the ECA treatment of planar flaws is equation. AIII [5] from which the S–N curve for the flaw is

$$S^m N = \frac{1}{C\, B^{\left\{\frac{m}{2}-1\right\}}} \int_{a_i/B}^{a_f/B} \frac{d(a/B)}{(Y\sqrt{\pi a/B})^m} \qquad [9]$$

Knowing the appropriate values of m and C, the solution for K, i.e. the function Y, and the initial and final crack lengths, the S–N curve can be calculated.

A simplified treatment of planar flaws is presented in BS PD 6493 which enables the limits on planar flaws to be readily related to the quality categories. The recommendations are based on calculations carried out using [9] applied to specific geometries for which the integral has been evaluated.

Using the procedure presented it is possible to relate quality category, (i.e. a particular S–N curve) to planar flaw size for embedded flaws of any shape in butt welds, and surface flaws of any shape in fillet welds and flush-ground butt welds. It is also possible to specify or estimate

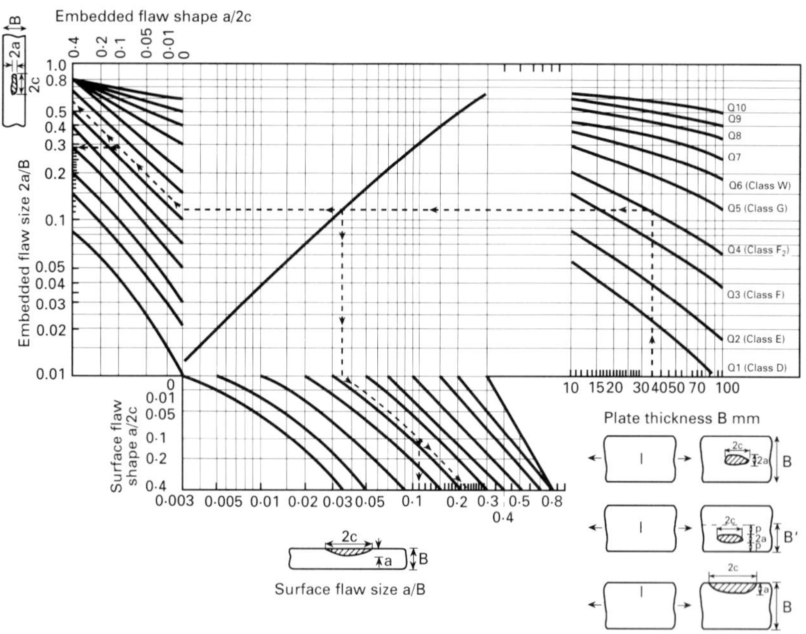

Fig. 29 (b) Flaws which can be assessed using Fig. 29 (a)

29 (a) Nomogram to determine allowable embedded or surface planar flaw sizes; (b) Flaws which can be assessed using 29a.

the critical crack size. Condensed presentations of the recommendations in BS PD 6493 are given in the form of nomograms, Fig. 29 and 30. Fig. 29 refers to surface flaws in plain material or flush-ground butt welds and embedded flaws in butt welds. The recommendations for embedded flaws refer specifically to flaws at mid-thickness. However, other embedded flaws can still be assessed, as the basis that they are in the middle of a thinner plate of effective thickness B′ = 2a + 2p, where p is the distance from the edge of the flaw to the nearest plate surface. Figure 30 refers to surface flaws at butt or fillet weld toes, that is cases when the flaw is situated in the field of stress concentration due to the welded joint geometry. Both nomograms can be applied to steel or aluminium alloy joints. Failure is defined as the attainment of through-thickness cracking, which was essentially the definition of failure in the fatigue tests used to determine the data upon which the main fatigue rules are based. To achieve simplicity, it has been necessary to introduce approximations into the nomograms, with the result that the allowable sizes of some flaws, particularly embedded flaws in butt welds and small surface flaws, may be underestimated using the nomograms, compared with assessments based directly on BS PD 6493. However, since embedded and small surface flaws are particularly difficult to size, this extra conservatism is not unreasonable.

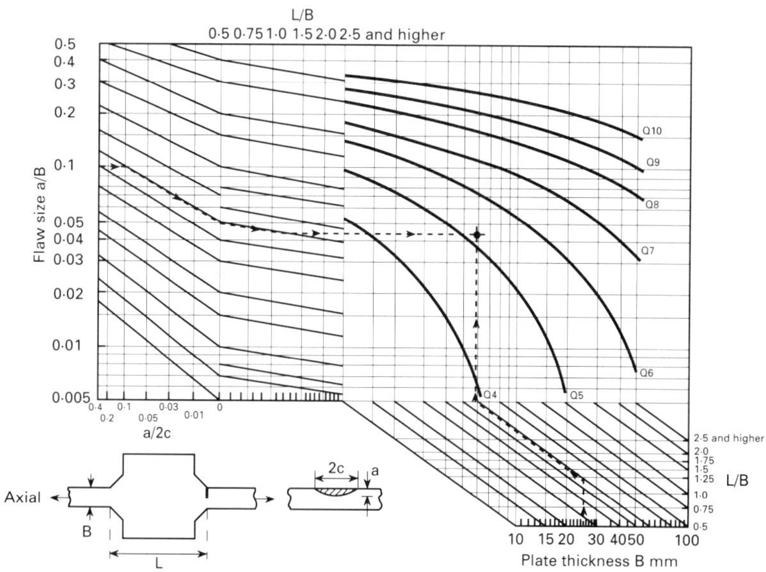

30 Nomogram to determine sizes of planar flaw at butt or fillet weld toes.

As an example of the application of the nomogram for butt welds, Fig. 29, consider its use, together with the information in Table 7, for the assessment of quality requirements for embedded flaws in an as-welded butt weld in the 35 mm thick flange of a steel girder, which experiences the same fatigue loading as a stiffener which is fillet welded to the flange. The welds attaching the stiffener have a lower fatigue strength than the highest fatigue strength which could be obtained from the butt weld (i.e. corresponding to Class E), and so they determine the highest quality category required. This corresponds to Class F and is therefore category Q3. However, because the stiffener is attached to a plate thicker than 22 mm, the effective quality category is lower. Referring to Table 4 and applying the thickness correction of $(\frac{22}{B})$ 0.25 to the stress range corresponding to Q3 reduces it to 165 N/mm². Hence the effective quality category is Q4 or Class F_2. From Table 7, the butt weld could contain a 35 mm long slag inclusion and up to 5% porosity. To use the nomogram, Fig. 29, enter the diagram at the plate thickness (broken line) and move on a vertical line to meet the required quality category. Movement horizontally to the embedded flaw axis enables the allowable depth of embedded flaw for different shapes to be read. In this example, the acceptable depths of embedded flaws at mid-thickness are between 4.2 and 20 mm depending on the flaw shape. Thus, for a shape a/2c = 0.1 $\frac{2a}{B}$ = 0.29, which corresponds to a flaw depth of 2a = 10 mm. If the butt weld was flush ground, Fig. 29 could also be used to determine acceptable surface breaking flaws (Fig. 30 would be used to assess surface flaws at the toe of the as-deposited butt weld). The procedure is to move vertically downwards from the diagonal line and to read off surface flaw depths for the required flaw shape in the same way as for embedded flaws. In the present example, acceptable flaws range from 1.2 to 8 mm in depth (e.g. a/B = 0.11, or a = 3.8 mm, for a/2c = 0.1).

The nomogram could be used in the reverse direction to determine the fatigue strength of a particular flaw or service crack.

In the case of assessments for flaws at fillet or butt weld toes it is necessary to take account of the attachment size (L) as well as plate thickness (B), since this influences the correction, M_k, discussed in Appendix III.

Suppose a 25 mm thick member with a 12 mm thick stiffener attached by 9 mm leg length fillet welds (i.e. L/B = 1.2) is found to contain a fatigue crack estimated to be 2.5 mm deep by 25 mm long (i.e. a = 2.5 mm. a/2c = 0.1). The member is required to have a fatigue strength corresponding to the Q6 S–N curve for fatigue service. Can the crack be left unrepaired? The routes taken on the nomogram, Fig. 30, are shown as a broken line. As will be seen, the horizontal and vertical lines meet at a point which falls between Q5 and Q6, indicating that the crack will not propagate through the plate thickness during the required life.

Recalling the discussion of the effect of plate thickness, the nomograms illustrate the situations in which an increase in plate thickness will lead to an increase or decrease in fatigue strength. In surface cracking in a butt weld without overfill there is no surface stress concentration, and so fatigue strength increases with plate thickness. In contrast, the curves for weld toe cracks show how the fatigue strength can decrease with increase in plate thickness.

Chapter 5

Improving the fatigue strength of welded joints

The previous chapters of Pt 2 have underlined that design against premature fatigue cracking requires the fitness of welded details to be matched with imposed stress levels and endurance requirements. Cost factors will obviously be influential in determining the degree of freedom in weld detailing: for example where the use of butt welds is necessary for the achievement of minimum overall scantlings. Conversely, the single-run fillet weld will often be attractive from the point of view of fabrication cost, given that it still satisfies what would normally be less demanding fatigue conditions.

Circumstances could exist, however, when it would be advantageous to be able to improve the fatigue performance of a particular kind of welded joint or detail. Methods for doing this are available. Each one requires a postwelding operation which, in turn, increases fabrication cost. Control and inspection of the operation are equally important considerations; in particular when the assumption has been made that the detail will be capable of withstanding more severe duty than in its original form. For these reasons, therefore, the use of improvement techniques tends to be somewhat limited in original construction, and rather more interest attaches to their value in the context of design modification or repair when, in general, the detail concerned will not subsequently be subjected to higher nominal stresses than before.

The techniques may also be considered to be more significant for lower strength details containing fillet welds. In the following descriptions reference is made to the treatment of details of this kind and fatigue results are typified by S–N data.

Reducing stress concentration effects

A major cause of fatigue damage in welded structures is the severe stress concentration at the toe of a fillet weld, where inherent crack-like flaws readily precipitate fatigue cracking. It follows that, if weld toes are machined or ground to eliminate such flaws and give a smoother profile, an improvement of fatigue strength should result. This is confirmed by experiment for transverse fillet welds treated by disc grinding and rotary burr machining, Fig. 31 and 32. The weld toes at the ends of short longitudinal fillet welds can also be improved by these methods, although in practice the techniques are more difficult to apply. Disc

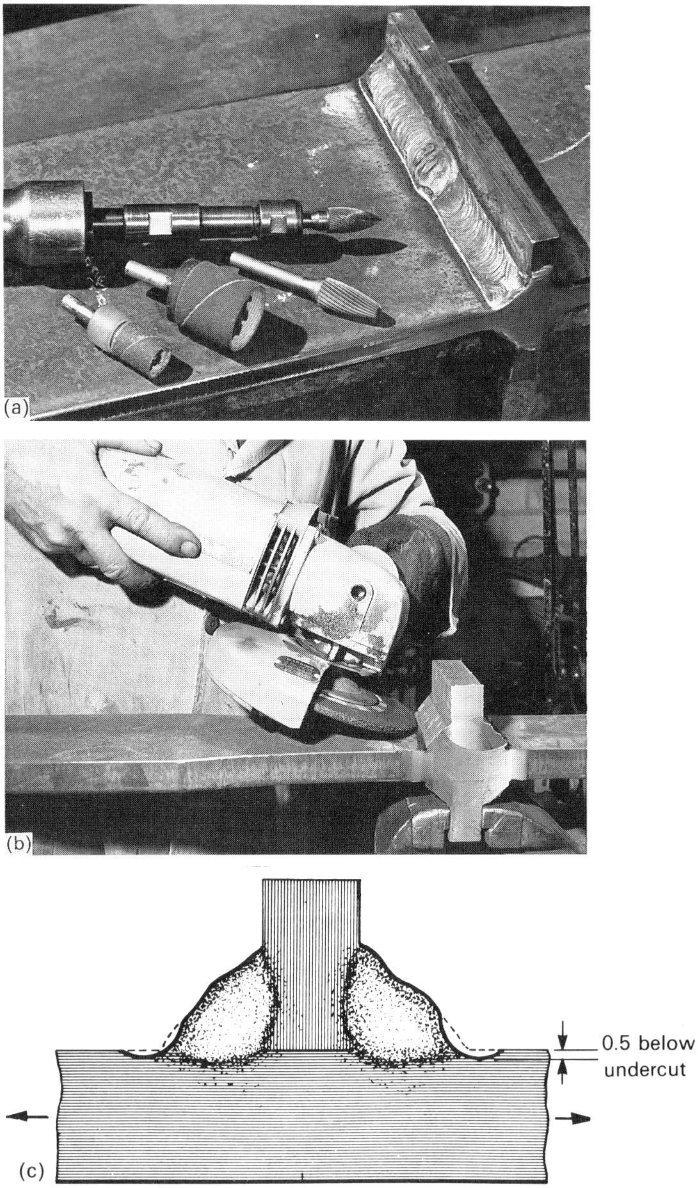

31 Weld toe dressing: (a) Tools used for burr machining and grinding; (b) Use of disc grinder; (c) Depth of grinding.

grinding is completed more quickly than machining with a hand-held rotary burr, although the latter is potentially more effective, Fig.32. In both, it is necessary to remove all traces of the original weld toe and material to a depth of at least 0.5 mm below any undercut, which in practice means that the depth will be of the order of 1 mm, as illustrated in Fig. 31c. Two other methods which appear to be as effective as local grinding but considerably faster are TIG and plasma dressing, Fig. 33. The principle employed is to remelt the weld toe and wash the weld pool into the plate surface in such a way as to produce a smoother weld profile, Fig. 33b; inherent flaws can be removed by remelting in this way. TIG dressing calls for precise positioning of the arc but the plasma technique, giving a larger area of heating, is less demanding in this respect, Fig. 34.

It will be clear that effective treatment of a fillet weld toe requires adequate access and careful control. Moreover, if joints are of a kind in which the weld root or embedded flaws could activate fatigue cracking, having dealt with the stress concentration at the toe, the full benefit of treatment may not be secured, depending on the fatigue strength of the embedded flaw.

It will be noted that the effect of the various weld toe dressing treatments is to rotate the S–N curve of the untreated joint to give a shallower slope and higher fatigue limit. Such a modification can be expected in view of the fact that the joint has been changed from one

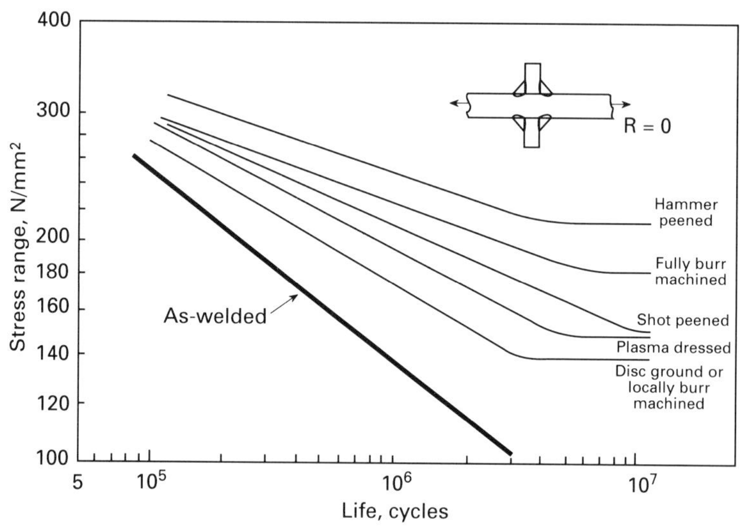

32 Typical improvement in fatigue strength of mild steel fillet welds resulting from toe dressing or peening.

in which the fatigue life was dominated by fatigue crack growth to one in which fatigue crack initiation occupies a significant proportion of the life. In other words, the welded joint now behaves more like an unwelded component. Although this has clear benefit, particularly in the low stress/high cycle regime where there is dramatic increase in the endurance limit, the joint now becomes unfavourably sensitive to factors which are less significant to crack growth than to crack initiation. In particular, corrosion pitting of the dressed surface virtually eliminates any benefit of dressing as a result of the reintroduction of severe stress concentrations. Also, the fatigue behaviour of the joint is now sensitive to applied stress ratio, to the extent that, again, the benefit of dressing is virtually lost if $R \geqslant 0.5$. On the positive side, however, the fatigue life of the dressed joint improves further for part-compressive applied stress, i.e. $R < 0$.

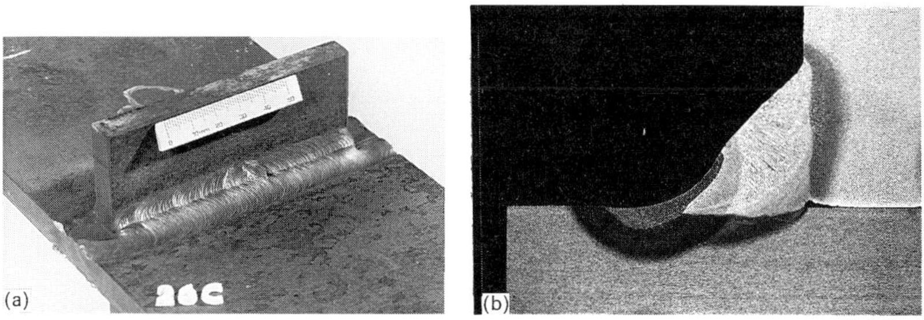

33 Plasma dressing of fillet weld toe: (a) Surface appearance; (b) Weld cross-section (×5).

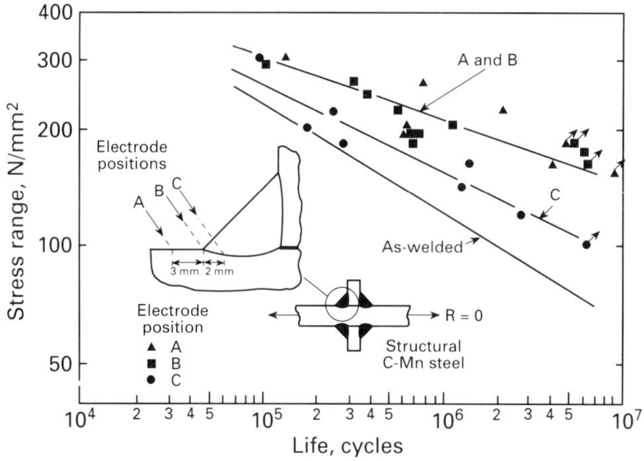

34 Influence of electrode position on fatigue strength of plasma-dressed fillet welds.

Introduction of compressive residual stresses

The effective stress produced when a component containing residual stresses is loaded is the sum of the applied and residual stresses. As has already been shown in Pt 1, applied compressive stresses become effectively tensile, and therefore damaging in fatigue, when they are superimposed on the high tensile residual stress present in an as-welded structure. Similarly, applied tensile stresses can become effectively compressive, and therefore less damaging, when superimposed on a high compressive residual stress, Fig. 35. Thus, if high compressive

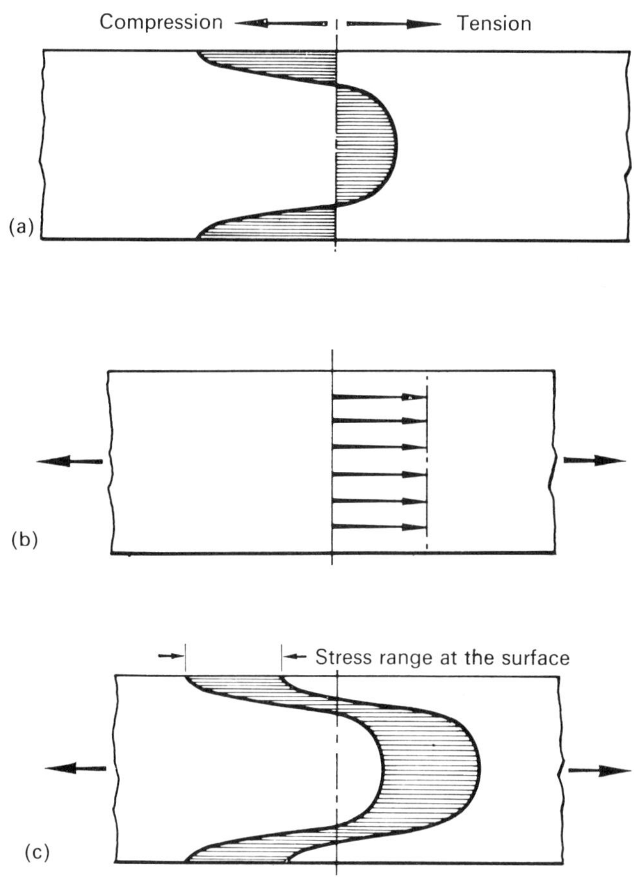

35 Superimposition of applied tensile and residual compressive stresses: (a) Residual stresses from mechanical working of surface; (b) Tensile stress from an applied load; (c) Resultant combined stress system.

residual stresses can be induced at the site of a stress concentration an improvement in fatigue behaviour will result.

Several techniques have been developed which are capable of inducing compressive residual stresses in a reasonably controlled manner, of which the most common involve cold working the surface. For example, shafts are shot peened and bolt threads cold rolled quite extensively as a means of improving fatigue performance.

Cold working the material surface produces a layer of compressive stress which is balanced by a residual tensile stress within the core of the material. As a result, the actual stress range at the surface under an applied tensile stress can remain predominantly compressive, depending upon the magnitude and extent of the residual stresses. The majority of fatigue fractures initiate at a surface so that treatment of this kind can secure a valuable increase in fatigue performance.

One way to introduce compressive residual stresses at the weld toe is by hammer peening, Fig. 36. This may be carried out using a pneumatic or electric hammer with a suitably round-ended tool. When applied to welded joints the residual stress system produced is more complicated than that described but still results in a region of compressive stress on the surface at the points of stress concentration. In general, for similar details, a slightly greater improvement in fatigue strength can be obtained by hammer peening than by dressing, Fig. 32, but again the actual improvement will depend upon the type of detail and the required life, being greatest at low stress/high endurances.

A disadvantage of hammer peening is the noise produced and, for some industrial environments, the technique would not be allowed for health reasons. An alternative technique which is quieter, easier to operate and potentially as effective is needle peening. Needle peening guns of the type commonly used to descale welds are ideal, Fig. 37.

A third method, which is well suited to the treatment of welded components in the workshop, is controlled shot peening. This technique is widely used to treat unwelded components and normally considerable care is necessary to produce consistent results and to ensure that the surface finish is not degraded by shot peening. However, experience indicates that the treatment of welded joints is far more tolerant to process variations and that the most important objective should be to maximise the deformation of the surface of the material. Then, as illustrated in Fig. 38, it is possible to achieve similar benefit from controlled shot peening as from hammer and needle peening.

Laboratory tests have demonstrated that two other techniques can also achieve the same objective of inducing compressive residual stresses, namely prior overloading and spot heating. Each has distinct limitations in the context of fatigue improvement of welded structures. Prior overloading, for example, requires the member concerned to be

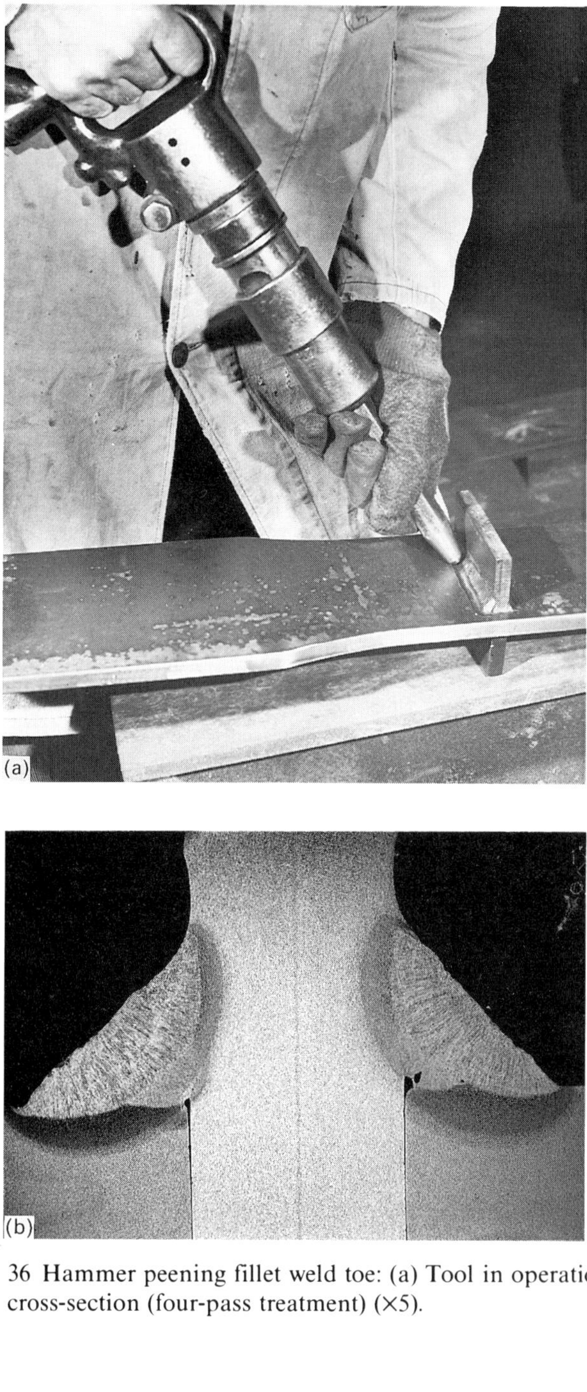

36 Hammer peening fillet weld toe: (a) Tool in operation; (b) Weld cross-section (four-pass treatment) (×5).

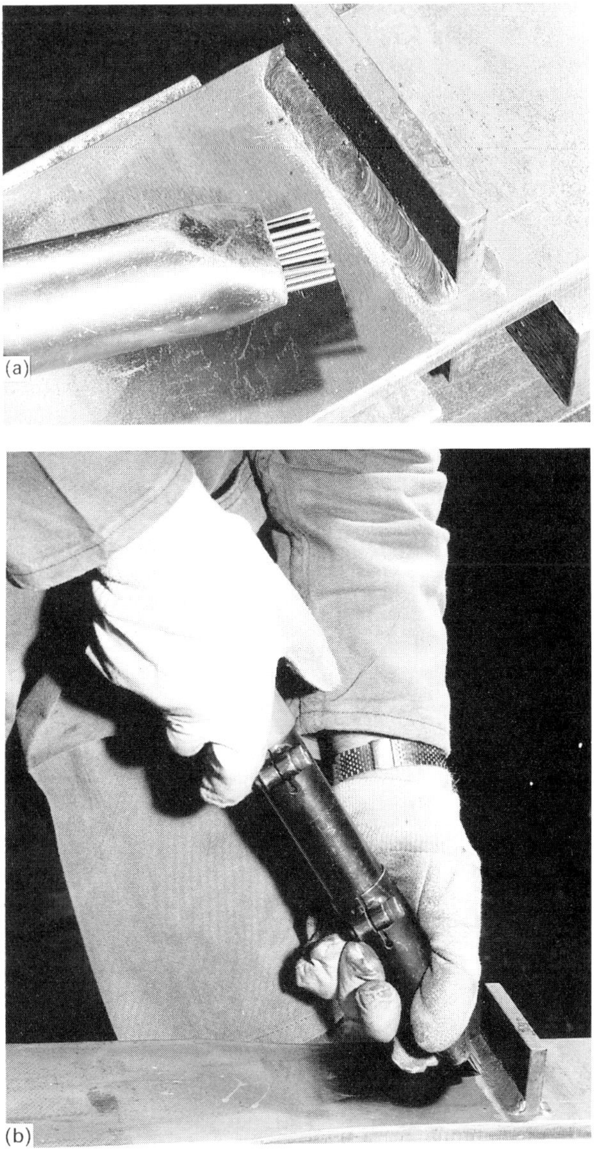

37 Needle peening fillet weld toes: (a) Tool and peened toe; (b) Peening operation.

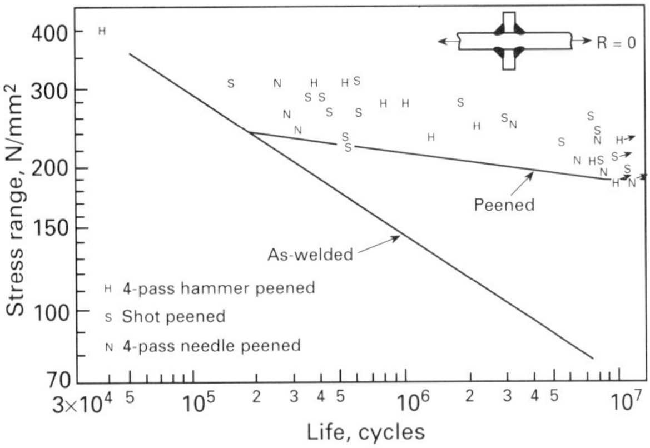

38 Comparison of fatigue life improvement due to hammer, needle and shot peening.

loaded not only to something in excess of the maximum service load but also in precisely the same way as the service load is applied. Beyond the proof testing of vessels, this is seldom practicable.

The principle involved in prior overloading is identical to that of tensile overloading, which produces crack growth retardation, in the context of fatigue under variable amplitude loading. By overloading to produce tensile stresses, all points of stress concentration, including flaws, are automatically located and treated to the benefit of subsequent fatigue behaviour. Even where possible, overloading a welded structure requires careful consideration of other aspects of behaviour such as instability or buckling, or even instable fracture; criteria such as these should be fully assessed. It is also emphasised that overloading is only beneficial if there are relatively severe geometric stress concentrations present, and they experience tensile stresses during the overload. Overloading of smooth, unnotched components can reduce fatigue strength, while compressive overstressing has virtually no effect on fatigue life.

The very different technique of spot heating requires, on the other hand, that the stress concentration points to be treated are identified beforehand. In spot heating, a small area near the stress concentration is heated in a rapid and controlled way to create conditions in which the heated zone experiences much the same sequence of restrained plastic deformation followed by the development of residual stresses as in welding. The resulting tensile residual stress acts radially from the centre of the heat spot while the compressive component is the tangential stress acting outside the spot area, Fig. 39(a). The heat spot must be positioned to ensure that the stress concentration being treated lies within the field of compressive residual stress. It will be apparent that, because of the limited extent of this field, only localised stress concentrations, like the ends of longitudinal attachments, Fig. 39(b) are amenable to the treatment; the toes of transverse welds cannot normally be treated effectively. The technique also offers a means of treating existing fatigue cracks, where it is used to introduce residual stresses which effectively clamp the crack tip closed and hence reduce, or even prevent, further crack growth. When the technique can be legitimately applied the resultant increase in fatigue life, especially at long endurances, can be substantial, Figs. 39(c) and 40.

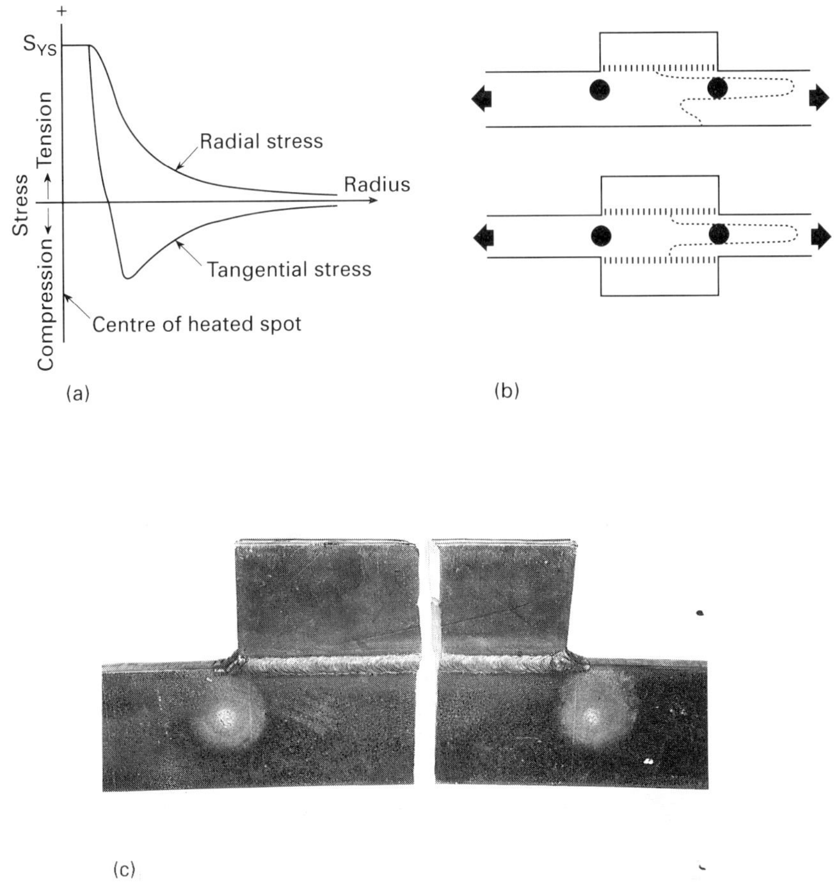

39 Spot heating weld ends: (a) Residual stress distribution; (b) Positions of heat spots for treating ends of longitudinal welds attaching gussets to plate edges; (c) Fatigue failure in plate with longitudinal edge gussets transferred from weld end (Class G) to middle of weld (Class B or C).

As the above techniques change the residual stresses in the region of the weld so that the high tensile residual stresses associated with the as-welded joint are replaced by compressive stresses, the fatigue strength of the joint is no longer independent of applied mean stress. Instead, the effective stress range (resulting from the superimposition of applied and residual stresses) increases as applied tensile mean stress increases, so that the benefit of the improvement method virtually

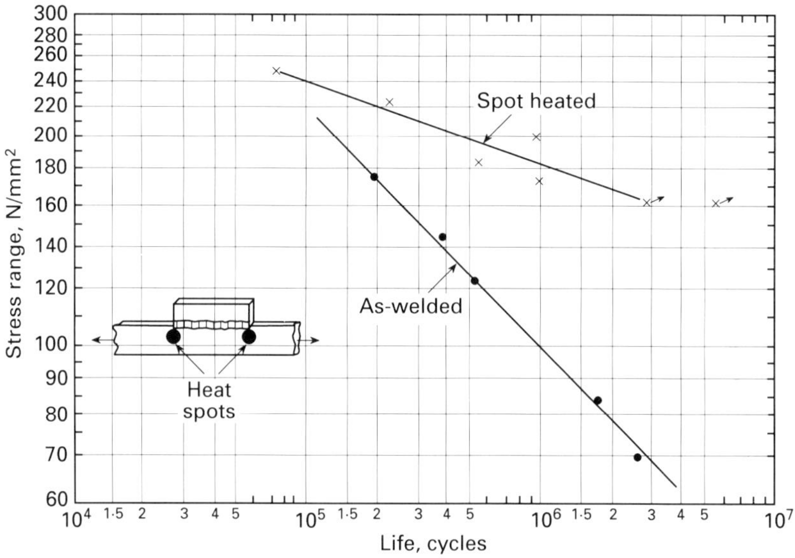

40 Effect of spot heating on fatigue strength of plate with edge attachment.

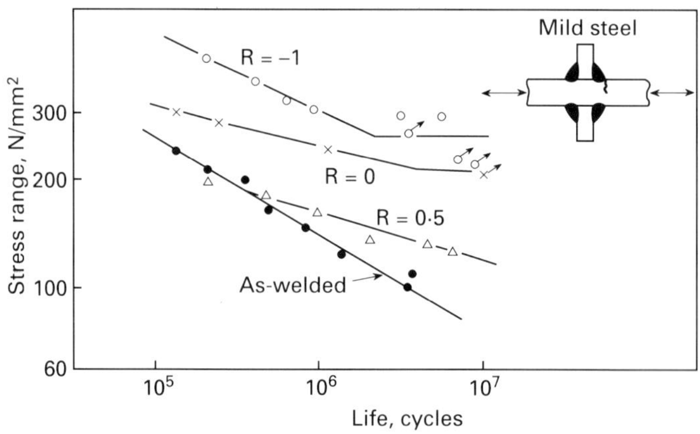

41 Effect of stress ratio on fatigue strength of hammer peened (O, X, δ) fillet welds; (●) as-welded.

disappears at high mean stresses, Fig. 41. However, in contrast, the benefits are particularly high if the joint is subjected to part-compressive loading, Fig. 41.

High strength steels

As has already been pointed out, the potential for using high strength steels to advantage in fatigue loaded welded structures increases with increase in the basic fatigue strength of the weld details, in that high design stresses are possible. Therefore, for some circumstances, it should be possible to use high strength steels to advantage if improvement techniques are applied.

Furthermore, there are some indications that the level of improvement due to an improvement technique will be greater in a high strength steel. In the case of weld toe dressing, a significant fatigue crack initiation period is introduced and, as discussed earlier, there is a tendency for the number of cycles required to initiate a fatigue crack to increase with material tensile strength. The residual stress techniques do not remove inherent welding flaws, but instead introduce benefit by reducing the rate of growth of those inherent flaws. However, there are some indications that, over a limited range of material strengths, the magnitude of the induced compressive residual stress increases with material yield strength, so that again the improvement technique is more effective in high strength steels. However, this finding is by no means universal and at present there are insufficient data available to produce general guidance on the increase of design stresses related to material strength for fatigue conditions.

Chapter 6

Typical service failures

The analysis of a service failure is invariably assisted by examination of fracture surfaces and possibly the microstructure of the material in the region of cracking. In this way it is possible to determine positively whether the crack has propagated by the fatigue mechanism or whether brittle fracture has intervened; some evidence of fatigue loading history may be gained or the deleterious influence of serious weld flaws revealed. Such observations can be very important to the planning of repairs, especially where premature failure has occurred and the need exists to restore adequate strength.

However, perhaps the greatest value of records of service fatigue failures comes from the convincing evidence they provide of poor or inadequate design, and hence the lessons that can be learnt.

Mechanical handling and earthmoving equipments are normally subjected to random loading amounting to many cycles in service and, not surprisingly, fatigue cracking is not uncommon. The first example illustrates a failure in a tractor shovel arm which initiated at a hydraulic ram attachment, Fig. 42. Welded-on brackets and attachments of this type are fairly common and can result in a relatively low fatigue strength. With careful design many of these fittings can be incorporated into the basic structure to eliminate the crack initiation points and possibly result in a more economic fabrication.

A similar detail was associated with fatigue cracking in a large fork lift mast, Fig. 43. Fatigue cracking initiated at the end of a fillet weld between the mast and a transverse connecting plate, Fig. 43(b), where a brittle fracture initiated from the small fatigue crack and demonstrated the need for steel with better fracture toughness for the masts.

Heavy transport vehicles with welded chassis components also contain many brackets, bearing pads, and similar details which can give rise to the type of failure shown, Fig. 44. It is sometimes difficult to eliminate such details from the design and it becomes necessary to reduce the stress levels by increasing the thickness of the component or to improve the behaviour of the detail by locally grinding the weld toes.

Chemical and processing plant usually contains a considerable array of pipework, steam heating coils, etc, Fig. 45 (a), which inevitably means a large number of different types of welded detail. These plants may appear to be completely static with regard to loading but considerable fluctuations in stress may be produced by changing pressure conditions, vibrations caused by peculiar flow conditions, and most particularly by temperature variations in the process. Figure 45(b) illustrates a pipe surrounded by a steam jacket where the variation in thermal stress was

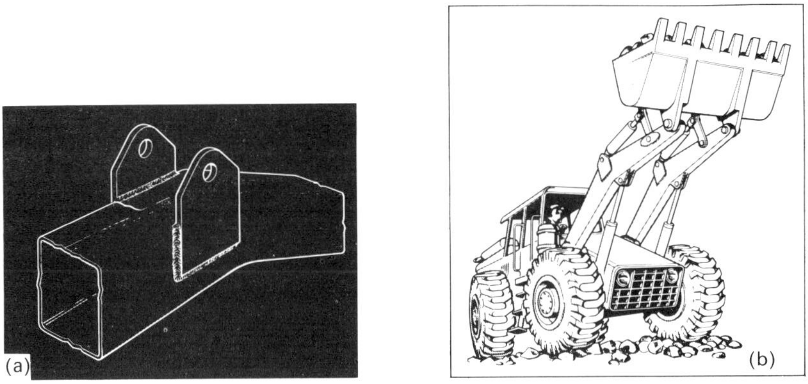

42 Fatigue failure of a tractor shovel arm: (a) Welded attachment at which fatigue crack initiated; (b) Failed member in operation.

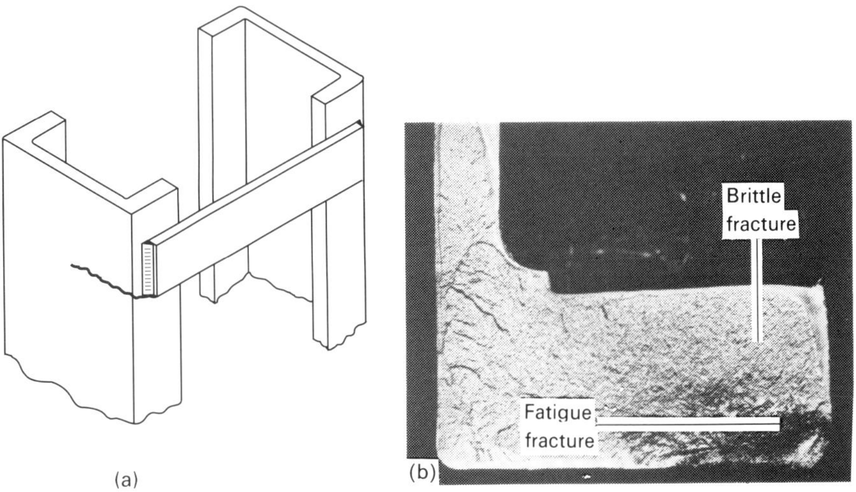

43 Failure of a fork lift mast: (a) Fillet welded connecting plate at which fatigue crack initiated; (b) Fracture surface showing brittle fracture which initiated from small fatigue crack at weld toe.

sufficient to produce failure from the fillet welded detail. Corrosion fatigue is also very common in this type of plant and it is essential to try to eliminate corrosion traps and crevices which would facilitate this type of failure.

The majority of fatigue failures in service are associated with fillet welds because of their lower intrinsic fatigue strength. Poorly made butt welds will behave in a similar fashion and it is important that correct information on joint preparation should be shown on the drawings. The example, Fig. 46(a), shows a failure of the top flange of a **bogie frame** which was fabricated in sections to facilitate forming, and then welded together before attachment to the side members. Figure 46(b) clearly shows the severe lack of penetration resulting from the use of an inadequate joint preparation which initiated the failure.

Many fabricated designs have been produced as direct copies of structures previously made in **cast steel**, Fig. 47(a). Although this may give a structure equal in rigidity to a casting, it also involves poor joint details which cannot tolerate the high stresses induced as a result of this rigidity. Winding drums, Fig. 47(b), and large pulleys are examples of this type of construction and Fig. 47(c) shows a typical failure in the radial web-to-hub joint made with load-carrying fillet welds on such a structure. In some situations the use of full penetration butt welds will give the necessary fatigue performance but often, by a more rational approach to the design, many details of this type can be eliminated altogether.

Power presses, Fig. 48(a), are perhaps a classic example of pulsating-tension loading and several failures have occurred, particularly at the frame corners on high duty machines. Figure 48(b) shows a fillet welded box section frame with cracks in the inside flange-to-web welds. Curved members of this type can give rise to poor fit-up between

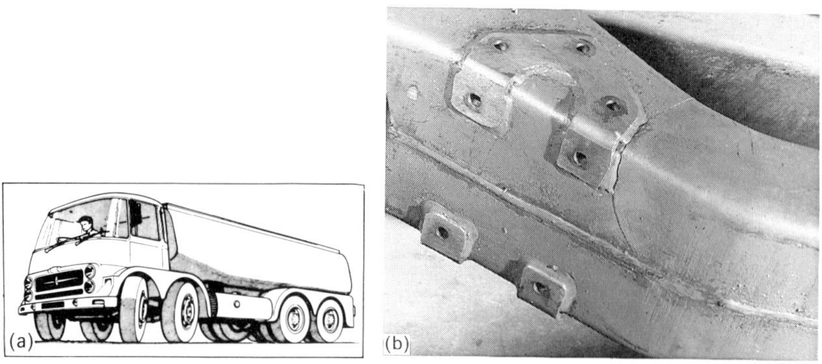

44 Fatigue cracking at fillet weld toe in a chassis.

the components before welding which results in a fillet weld with limited throat thickness. The problem is aggravated by the local lateral bending stresses in the flanges induced by their curvature. Full penetration butt welds made on a backing strip in place of the fillet weld will overcome the problem of poor fit-up and will also be more resistant to local bending stresses.

Fatigue cracking often occurs in **rotating equipment** since, normally, many cycles of loading are accumulated in service. Failure of this kind

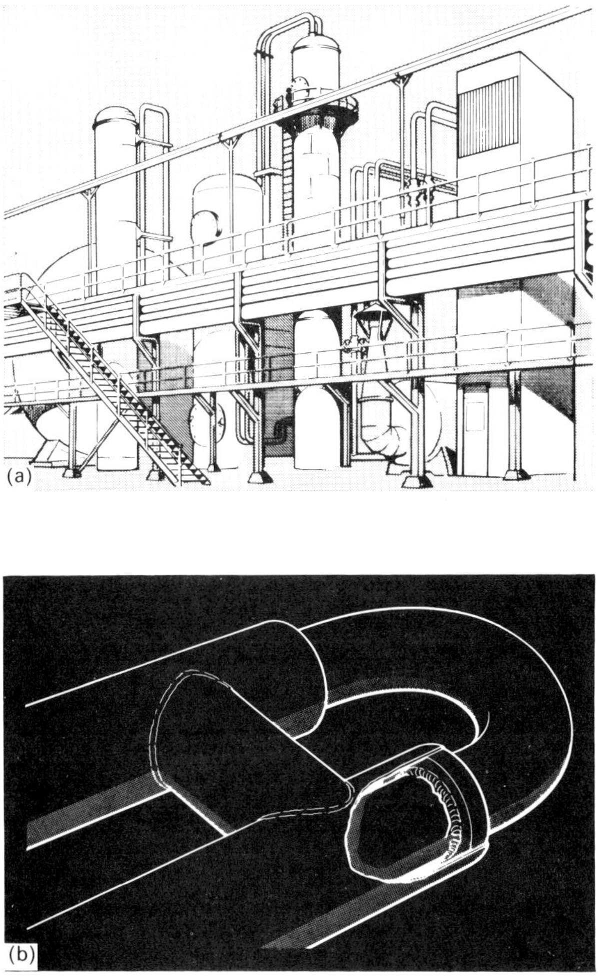

45 (a) Welded pipework in processing plant; (b) Fatigue crack at fillet weld toe due to thermal fatigue.

can be particularly dangerous in **impellers** and Fig. 49 shows just one of the possible sites for cracking in a welded impeller. Impeller blades are balanced, to reduce the out-of-balance forces which are one of the sources of cyclic load during operation, by attaching weights. In the example shown it is especially unfortunate that fatigue cracking initiated at the toe of the weld attaching such a weight. In equipment such as impellers where there are many sources of cyclic loading and the calculation of their magnitude is difficult, it is particularly important to ensure that the best possible details are selected and that attachment welds are avoided.

It is often convenient to make welded attachments directly to **shafts** for various types of machinery, motor armatures, etc, Fig. 50(a). Figure 50(b) shows the result of such an attachment. The service life of rotating components often involves several million cycles of stress fluctuation which means that the level of stress must be kept very low to avoid such failures. Machining or grinding subsequent to welding may be used to advantage but care must still be exercised in selecting the right type of detail and appropriate stress level.

The potential benefit to be gained from the machining of welds can, however, be critically dependent upon weld quality, notably when long cyclic lives are required. Figure 51 shows fatigue cracking which initiated from a defect in the last pass of a flange attachment weld.

Harmful defects might also arise when worn shafts are reclaimed by depositing weld metal and remachining, Fig. 52. In this respect it is particularly important to note that the original shaft material will not usually have been selected to be readily weldable, and many examples

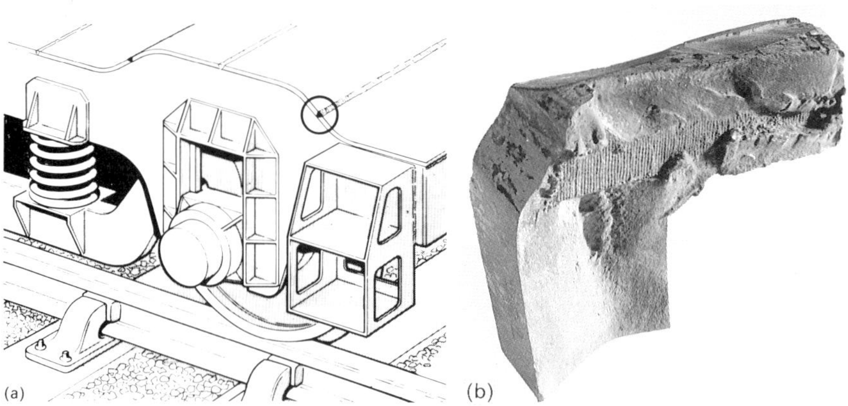

46 Fatigue failure in welded bogie: (a) Butt weld which failed; (b) Fracture surface showing fatigue crack which initiated from region of incomplete penetration.

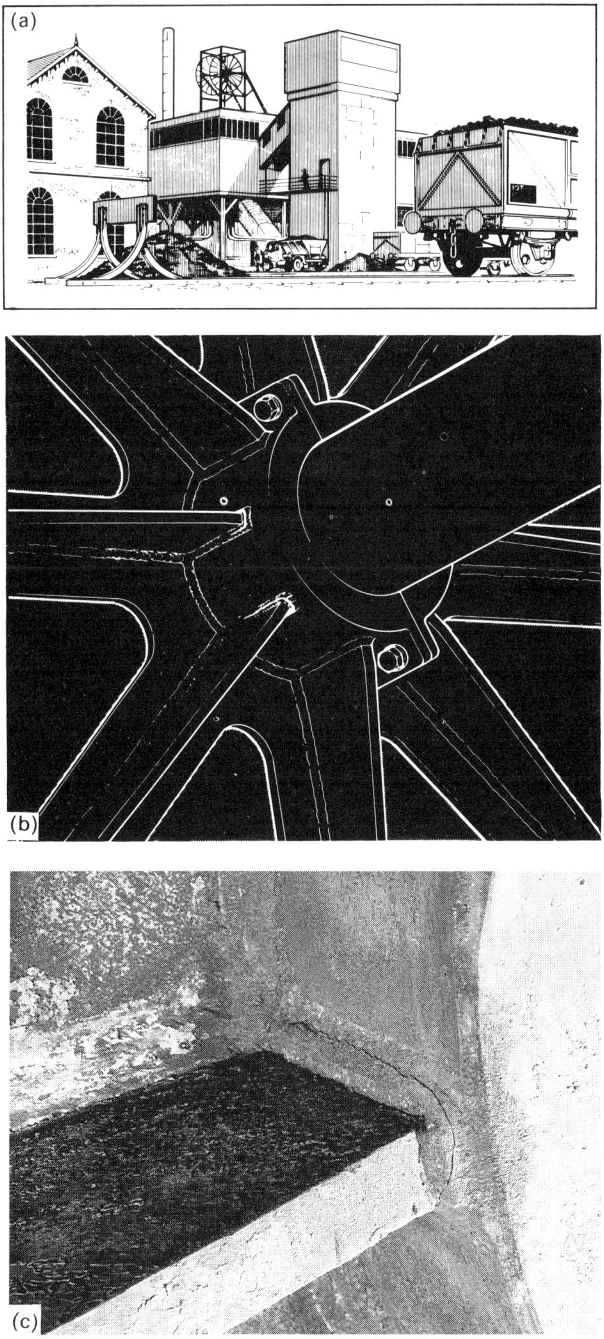

47(a) Welded 'copy' of cast structure: (b) Mine winding drum; (c) Fatigue cracking in throat of fillet weld in mine winding drum.

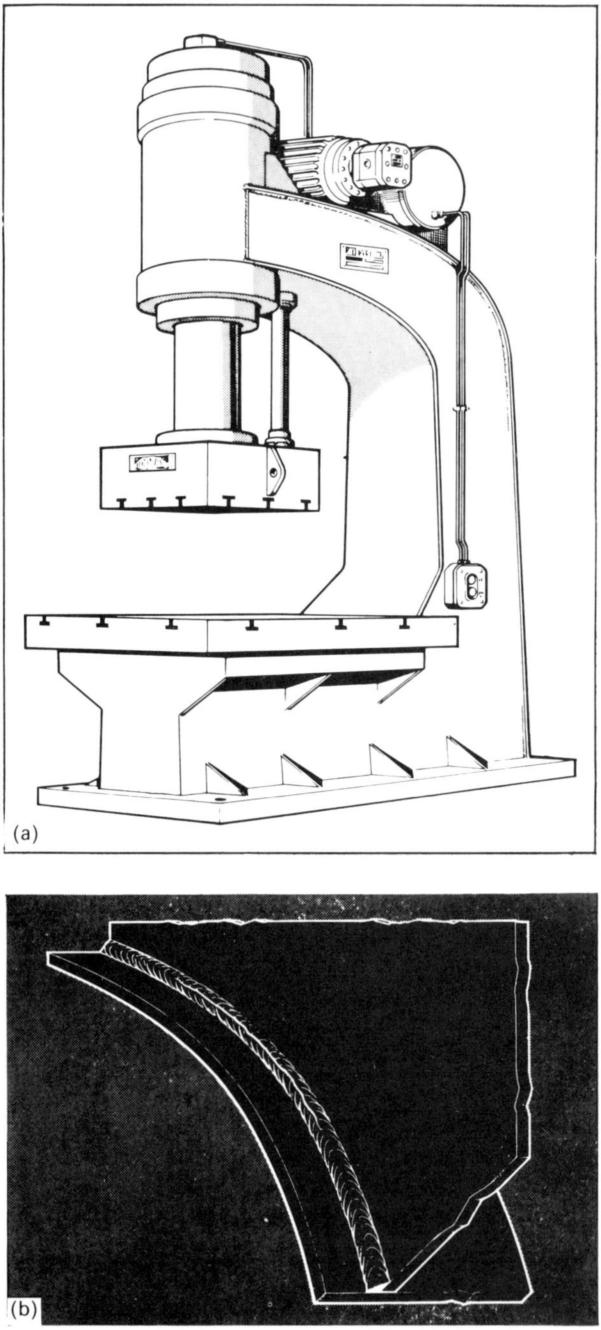

48 (a) Power press; (b) Fatigue cracking in throat of web-flange fillet-weld.

are on record where reclaimed shafts in high carbon equivalent steel have developed fatigue crack growth from heat affected zone, hydrogen-induced cracks situated below the machined surface.

In general, the welding of a previously unwelded component, either for repair or reclamation, as in the case of the shaft, must be considered with caution. Fatigue design stresses in unwelded material can be much greater than those which would be sufficient to cause fatigue failure from a weld, and consequently the fatigue life can now be expected to be much less than that of the original component.

The arduous conditions experienced by tubular offshore structures in the North Sea has caused many cases of fatigue cracking and sometimes failure of major members, or even the whole structure as in the case of the Alexander Keilland. The most vulnerable connection is that between brace and chord members. Figure 53 illustrates the two possible modes of failure, the chord failure resulting from local bending of the chord wall under loading from the brace. The cases shown occurred because the brace member was situated close to the sea surface where it experienced fluctuating buoyancy loads; as a result of this and similar cases, the practice of positioning braces in this zone ceased. However, similar fatigue cracks have occurred in other regions, highlighting the general problem of forecasting the severity of the wave loading conditions.

The ease with which parts can be rapidly joined by welding means that it is widely used to provide **fabrication aids**, some temporary but some permanent. The consequences of such welds in fatigue-loaded structures can be disastrous. Often the designer will not even be aware of the presence of the welds and the very fact that they are considered by the fabricator to be non-structural means their quality may be poor.

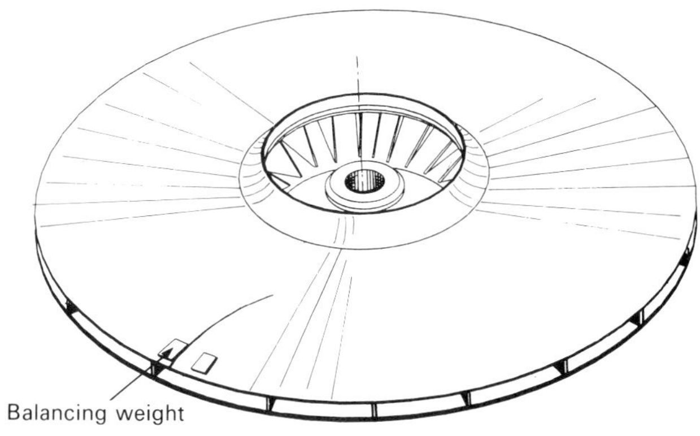

49 Fatigue cracking in welded impeller.

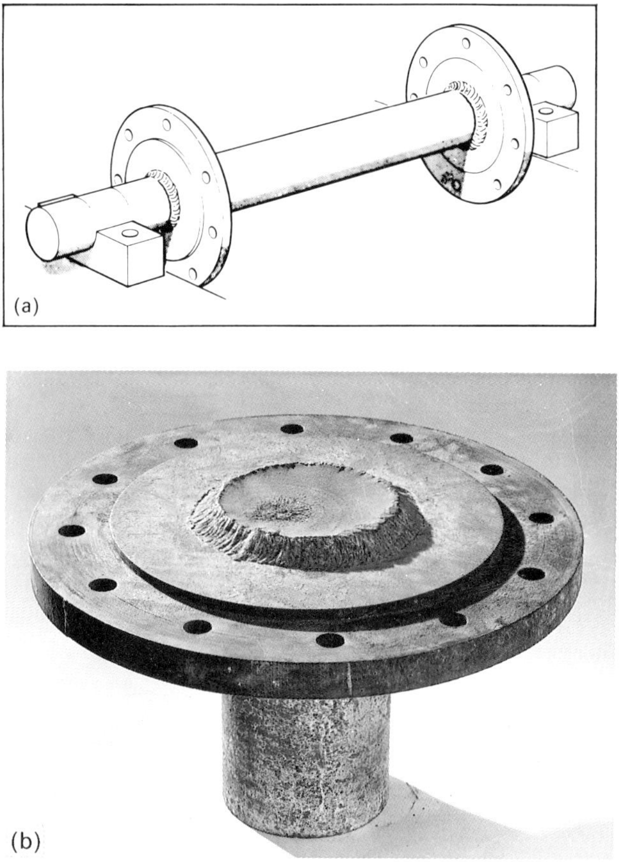

50 Fatigue failure of a welded shaft: (a) Shaft with welded flanges; (b) Fatigue failure which initiated at fillet weld toe.

A number of examples of fatigue failures which have originated from welded fabrication aids are shown in Fig. 54.

The first example, Fig. 54(a), concerned a backing strip provided to enable a full penetration butt weld to be made in a tubular member from one side only. However, the strip was made up of several pieces welded together. The welds, which clearly were not considered to be significant, were only partially penetrated. Although the backing strip was not required structurally, the very fact that it was welded to the tube meant that it became an integral part and attracted loading. As a

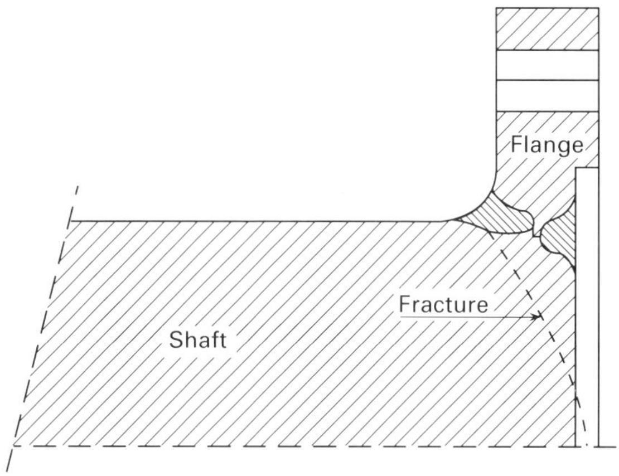

(a)

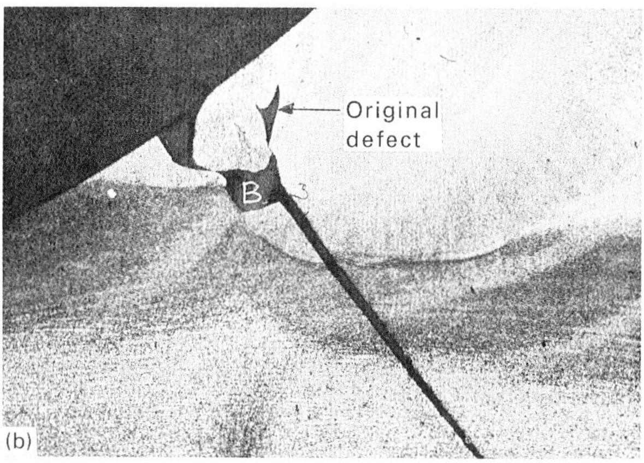

(b)

51 Service fatigue failure in shaft-to-flange weld which initiated from welding defect (arrowed) (region A was distorted and material from B was detached as a result of failure) (×10).

result, fatigue cracking initiated from the lack of penetration flaw and propagated through the main butt weld, causing the member to fail. It was recognition of this type of potential problem, and the risk of cracking when welding over the space between backing strips, which led to the requirement in the design classification system (see Chapter 1) that short lengths of backing strip should be joined with full penetration butt welds.

The second example, Fig. 54(b), also concerned a tube this time a large-scale brace member in a fixed offshore structure. In this case the fabrication aid was an elliptical opening in the brace to allow access to the inside to facilitate welding and inspection of the butt welds at the ends, where the brace met main leg members. The opening was finally closed by welding back the piece originally removed. Due to poor fit-up and an unfavourable welding position the resulting weld contained flaws, from which fatigue cracks initiated in service. A number of members were affected and repair proved to be extremely costly.

The final example, Fig. 54(c), occurred in an orthotropic suspension bridge crossing a river. The bridge deck, which was of trapezoidal section, was made up from 18 m lengths which were floated to the site prior to erection. Buoyancy was achieved by welding ends to the deck sections, which were subsequently removed when the bridge deck was complete. However, the actual welds were not removed and they remained as fillet welded attachments to the undersides of the main plate stiffeners. Fatigue cracks initiated from the weld toes in the stiffeners under the traffic-induced fatigue loading.

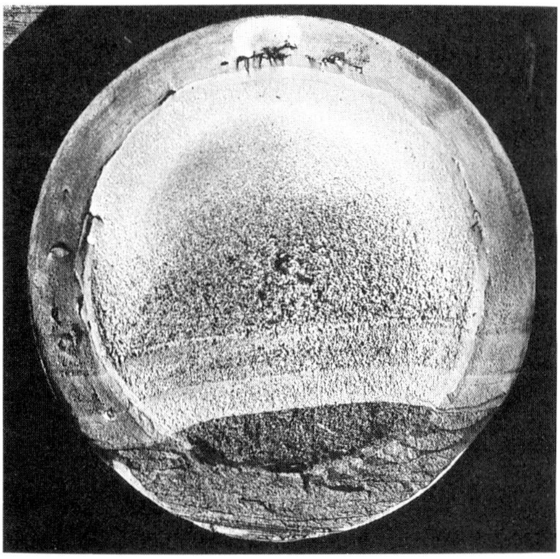

52 Fatigue failure in rebuilt shaft which has initiated from welding defects.

All these examples highlight the importance of *any* weld in a fatigue loaded component or structure and the need to ensure adequate weld quality even in secondary welds if they subsequently form part of the main structure. Designers in particular should be aware of the possibility that additional welding might be performed and, in fatigue-critical structures, ensure that no unauthorised welding is carried out without their knowledge.

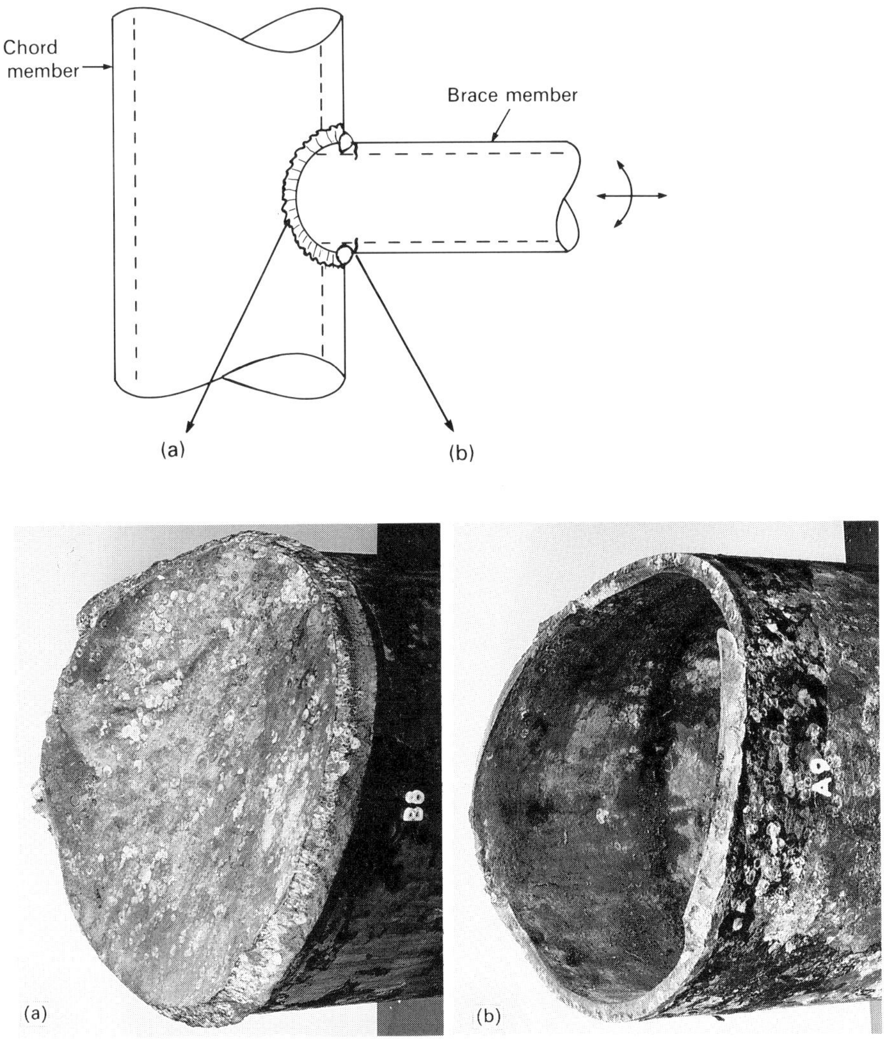

53 Modes of fatigue cracking in tubular joint: (a) From weld toe into chord; (b) From weld toe into brace.

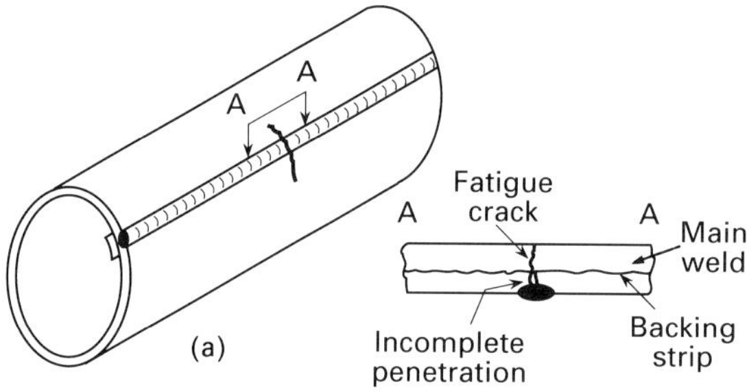

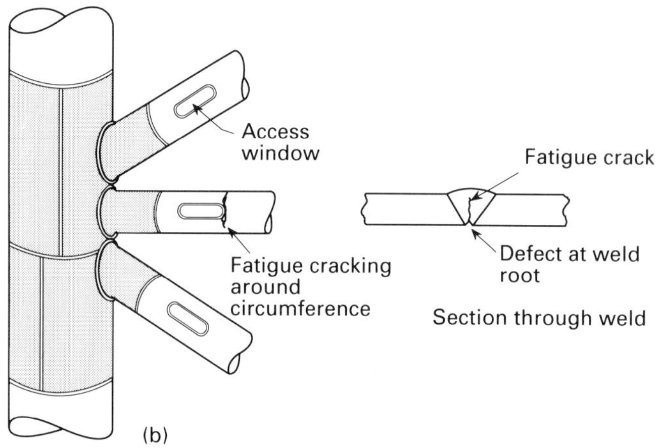

54 Fatigue cracking from: (a) Incomplete penetration defect in butt weld in backing strip; (b) Brace member in off shore structure from defect at root of single-sided closure weld (shaded area shows node); (c) Toe of a fillet welded fabrication aid.

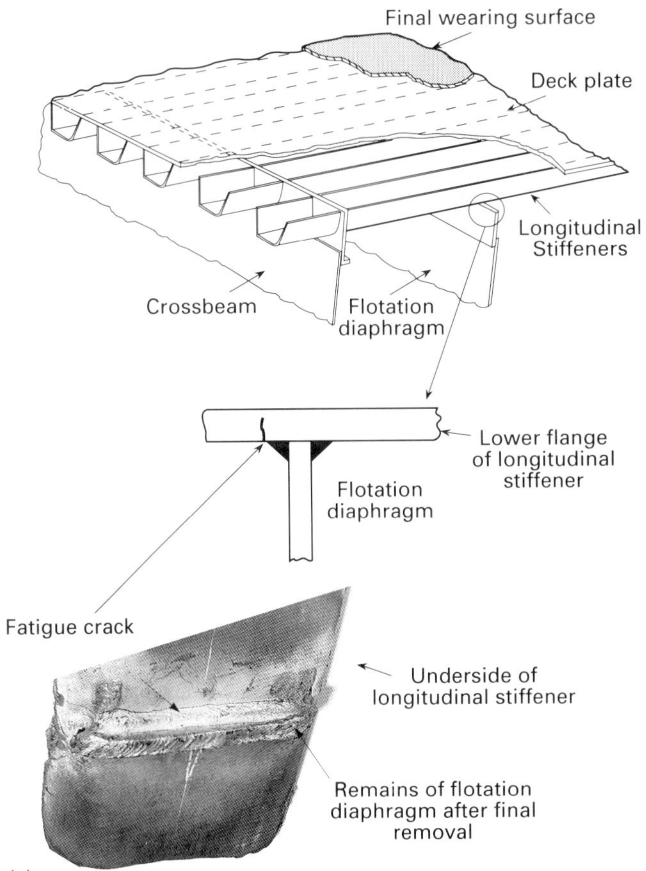
Final wearing surface
Deck plate
Longitudinal
Stiffeners
Crossbeam
Flotation
diaphragm
Lower flange
of longitudinal
stiffener
Flotation
diaphragm
Fatigue crack
Underside of
longitudinal stiffener
Remains of flotation
diaphragm after final
removal

(c)

Chapter 7

Conclusions

Reference to cases of service fatigue failure may seem a rather negative way to close a book which aims to help designers to avoid such failures. However, they are surely the best indicator of where problems can occur. It is remarkable how few service fatigue failures provide 'surprises' and how few problems occur when close attention is paid to fatigue design. Of course, problems still arise because of uncertainty over loading, including abuse in service and unexpected sources of fatigue loading, and the presence of weld imperfections, but in general following the design recommendations produces welded structures with satisfactory fatigue lives. Future developments will include the provision of design data for welding processes not presently covered, better guidance on the treatment of complex loading, advice on the relation between the design data and the results of detailed stress analysis (e.g. finite element or strain measurement), perhaps using the hot spot concept currently applied to tubular joints, a better integration of fatigue life prediction methods (e.g. fracture mechanics) into the design process and a greater availability of characteristic service load spectra.

Meanwhile, the following checklist is offered to designers as a reminder of some of the potential pitfalls:

- Consider all possible sources of fatigue loading and estimate load spectra as accurately as possible.
- Assess all welds as potential sites for fatigue cracking.
- Choose the better class weld details, but if this is not possible locate lower class details in areas of low stress.
- Avoid stress concentrations due to structural discontinuities by ensuring smooth load paths and gradual changes in stiffness.
- Consider the possibility that additional welds (e.g. casual attachments, fabrication aids) may arise, and, if necessary, prohibit unauthorised welding.
- Provide the fabricator with adequate access to facilitate welding.
- Avoid setting unnecessarily stringent quality requirements if they

are not justifiable, with the option to apply fitness-for-purpose criteria for assessing the acceptance of imperfections.

- Communicate specific weld quality requirements to the fabricator and inspector.

Appendix I

Cumulative damage calculations using Miner's rule

In cumulative damage calculations using Miner's rule stress ranges below the constant amplitude fatigue limit, S_0, are assumed to be non-damaging. However in practice they become damaging as crack growth occurs under higher stress ranges in the spectrum. The omission of the damaging effect of these low stresses can lead to serious under-estimates of resulting fatigue life, as there are many practical load spectra in which the majority of cycles produce relatively low stresses. Possible ways of dealing with this problem are as follows:

(a) The simplest approach is to assume that there is no fatigue limit and to extend the S–N curve downwards *ad infinitum*. Depending upon the stress spectrum this may be practical, but often the damaging effect of low stresses is excessively overestimated.

(b) Independently of the design rules, fracture mechanics could be used to calculate the life on the basis that stresses will be damaging if the corresponding stress intensity factor range, ΔK (which depends upon both stress range and crack size), exceeds the threshold value (Appendix III).

(c) Use a modified S–N curve at low stress ranges which, when used with Miner's rule, predicts the same fatigue damage as (b). For joints in air it has been found that extrapolation of the S–N curve below S_0 (that is beyond 10^7 cycles) with a slope of $m + 2$ would be suitable for a wide range of spectra. Alternatively, the curve could be extrapolated at the same slope to a lower effective fatigue limit, S'_0, corresponding to an endurance of 2×10^7 cycles. Both approaches have been adopted in design codes, although the former is generally favoured and, in time, may be the method adopted in all Codes. Details of the equations of the modified S–N curves, which have the form $S^{m+2} N = B$, and values of S'_0 for 2.3% probability of failure are included in Table AI.1.

Cumulative damage calculations are simplified if the stress (or load) spectrum is expressed in terms of proportions, p_i, of the maximum value and the sequence is arranged in decreasing order of magnitude.

Thus

$$S_i = p_i \, S_{max} \qquad \text{[AI.1]}$$

From [5] Miner's rule can then be expressed as

$$S_{max}{}^{m} \, \Sigma p_i{}^{m} n_i \ngtr A \qquad \text{[AI.2]}$$

Example 1: Determination of fatigue life

The need for this analysis is most likely to arise when a check is required on the fatigue life of an existing structure. The maximum stress range and service stress spectrum over a given time period, for example (n_i, $p_i \, S_{max}$ values over a year), are known. Direct application of [AI.2] with the assumption that at stress ranges below S_0 the modified S-N curve

$$S^{m+2} \, N = B \qquad \text{[AI.3]}$$

applies, gives

$$q \left[\underbrace{\frac{S^{m}_{max} \, \Sigma p_i{}^{m} n_i}{A}}_{\text{stresses above } S_0} + \underbrace{\frac{S^{m+2}_{max} \, \Sigma p_i{}^{m+2} n_i}{B}}_{\text{stresses below } S_0} \right] = 1 \qquad \text{[AI.4]}$$

where q is the life in years.

Table AI.1 Proposals for treatment of low stresses in cumulative damage calculations (2.3% probability of failure)

Class	Modified endurance limit at 2×10^7 cycles, S'_0, N/mm²	Modified S-N curve S^{m+2} N = B beyond 10^7 cycles: m + 2	B
B	84	6	1.00×10^{19}
C	64	5.5	2.55×10^{17}
D	42	5.0	4.18×10^{15}
E	37	5.0	2.29×10^{15}
F	32	5.0	1.02×10^{15}
F_2	28	5.0	5.25×10^{14}
G	23	5.0	2.05×10^{14}
W	20	5.0	9.77×10^{13}

Example 2: Determination of allowable stress

This is the calculation which a designer is most likely to require. In this case the service load spectrum over a given time period (for example n_i, p_i W_{max} values over a year) and the required life are known. Also, it will be assumed that the member under consideration is subjected to axial loading, so that the stress range is given simply by $S = \frac{W}{X}$, where X is the cross-sectional area of the member. However, similar calculations can be performed if the loading produces bending, using $S = \frac{M}{Z}$, where M is the bending moment and Z is the section modulus.

The calculation is more complex than in the previous example because the cycles which correspond to stress ranges below S_0 are not known in advance. Applying Miner's rule gives

$$q \left(\frac{W_{max}}{X}\right)^m \qquad \Sigma\, p_i{}^m n_i \qquad = \text{A at failure} \qquad [\text{AI.5}]$$

This equation is evaluated for all cycles in the spectrum to give a value of X and hence S_{max}. The calculation could stop here, in which case a conservative estimate of S_{max} will be obtained. To reduce the required section further, S_0/S_{max} is calculated to give P_0 corresponding to S_0. The evaluation of Miner's rule is then repeated, as follows:

$$q \left[\underbrace{\frac{(W_{max}/X)^m \Sigma p_i{}^m n_i}{A}}_{p_i > p_0} + \underbrace{\frac{(W_{max}/X)^{m+2} \Sigma p_i{}^{m+2} n_i}{B}}_{p_i < p_0} \right] = 1 \qquad [\text{AI.6}]$$

Solving this equation, possibly by trial and error, gives a second estimate of X ($X_2 < X_1$). The procedure is repeated until successive values of X are virtually identical or X corresponds to some preferred plate thickness.

Appendix II

Cycle counting

Cycle counting is the procedure which is carried out to convert a stress or strain spectrum in the form of a complex waveform into a sequence of identifiable cycles to enable Miner's rule to be used to calculate the fatigue damage produced by the spectrum. Cycles are easily defined when the spectrum is characterised by constant minimum stress. The question of what constitutes a cycle when the minimum level varies presents a more complex problem, and a number of analytical and summation methods have been evolved to deal with this situation.

Three counting techniques are described here which are known respectively as the **range-pair**, **rainflow**, and **reservoir** methods. These are regarded as most relevant to the field of welded joint design by virtue of their aim of producing a summation of stress ranges. Range-pair and rainflow counting achieve this by combining positive and negative half-cycles, but differ in the way in which they define half-cycles; the reservoir method, on the other hand, employs a technique which identifies complete cycles in terms of stress range.

The range-pair counting method

This method simulates the action of the Vickers strain counter, a device which can be attached to structures such as aircraft to count strain cycles experienced over a period of operation at a particular location, and its application can best be described in terms of the action of the counter.

In general terms, separate records of preselected stress ranges are made which, in practice, correspond with equal intervals of stress taken from a datum which coincides with the minimum applied stress. The interval is a counting dead zone since stress variations of less than one interval are not recorded.

Each peak is regarded as defining a positive range acting from the datum. Thus, as stress increases to a peak, all counters between the datum and the peak not already activated are activated (or, in effect, a record is made, as entered in column A, Fig. AII.1). Excursion 0–A in the Figure therefore activates all counters between the datum and A; similarly, D–E activates those counters between the datum and E. As stress decreases from the peak to a trough, negative stress range is measured from the peak. Counters within this range, starting from the datum, are triggered (entry in column T) to complete a count; excursion A–B, for example, triggers the first three counters. It follows that, when B–C is applied, only the first three counters between the datum and peak C are not activated, but become so by this particular excursion.

It will be noted that the counters take no account of mean stress: any peak-to-trough change of, say, 100 N/mm^2 triggers counter numbers 1 to 5, given that the selected interval is 20 N/mm^2. Moreover it is evident that, when a peak or trough does not coincide with a selected stress level, the range is underestimated by an amount which can be as much as the counting dead zone. Clearly, for the greatest accuracy, the interval between stress levels should be as small as is practicable.

Finally, when a trough coincides with the datum, all activated counters are triggered. This occurs at locations H and P.

Details of the whole count are given in the Table in Fig.AII.1. The number of cycles corresponding to each stress range is then determined by subtracting the number of counts made at the **next highest** stress range from the number of counts made at the range in question, as indicated.

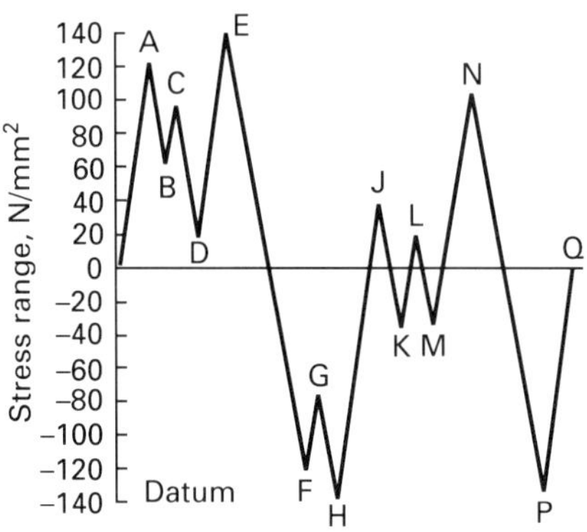

Stress range N/mm^2	A	T	A	T	A	T	A	T	A	T	A	T	A	T	A	Total	Total cycles
280					×			✓								1	1
260	×					✓										1	0
240	×					✓							×	✓		2	1
220	×					✓							×	✓		2	0
200	×					✓							×	✓		2	0
180	×					✓							×	✓		2	0
160	×					✓			×					✓		2	0
140	×					✓			×					✓	×	$2\frac{1}{2}$	$\frac{1}{2}$
120	×					✓			×					✓	×	$2\frac{1}{2}$	0
100	×					✓			×					✓	×	$2\frac{1}{2}$	0
80	×					✓			×					✓	×	$2\frac{1}{2}$	0
60	×	✓	×	✓	×	✓	×	✓	×	✓	×	✓	×	✓	×	$7\frac{1}{2}$	5
40	×	✓	×	✓	×	✓	×	✓	×	✓	×	✓	×	✓	×	$7\frac{1}{2}$	0
20	×	✓	×	✓	×	✓	×	✓	×	✓	×	✓	×	✓	×	$7\frac{1}{2}$	0

Datum O A B C D E F G H J K L M N P Q

AII.1 Range-pair counting method.

The rainflow counting method

In this method the waveform is imagined to be a series of rooftops down which rain can flow. It is viewed with the time axis in the vertical position, the starting point being at the top. The rooftop image is produced by representing the stress fluctuations as a triangular waveform. Rules are imposed on the way in which rain can flow so that the waveform is broken down into a number of rainflow paths. The beginning and end of each path define the extremities of a half-cycle.

Taking the same spectrum as before, Fig.AII.2, rainflow starts at the origin, 0, and is allowed to continue down the first roof until a peak or trough is reached. At this point the rain falls vertically until it reaches another roof, down which it continues. Rainflow from left to right and right to left is considered. Rainflow continues until one of the following conditions arises:

- Rain flowing down a rooftop encounters rain falling from a roof above. Rainflow from E to G stops at F for this reason.
- Rain falling vertically passes opposite a

  (peak which is more positive) than the (peak)
  (trough which is more negative) than the (trough)

  start of the rainflow path under consideration. The origin is regarded as a trough if the stress proceeds in a positive direction. Thus, flow path 0–A–F–G stops after G because the trough at H is more negative than that at 0. Similarly, path A–B–E stops because the peak at G is more positive than that at A.
- The end of the waveform is reached. Thus, unfinished flow paths G–H–K–L, L–M–S–T, and U–V would all be considered to be complete at the end of the analysis.

Each complete flow path is regarded as a half-cycle and half-cycles of equal range are combined to give a complete cycle. Actual ranges may be approximated to multiples of a selected stress interval to limit the number of stress ranges considered. In the example, intervals of 20 N/mm^2 have been used. The results of the rainflow count are shown in Fig.AII.2.

The reservoir counting method

An alternative to rainflow counting which is particularly suitable for short waveforms is the reservoir counting method. This is the method recommended in the British Standard bridge design code for dealing with the short stress histories produced by individual loading events. It differs from the other methods considered in that complete cycles in terms of the stress range are produced, rather than half-cycles.

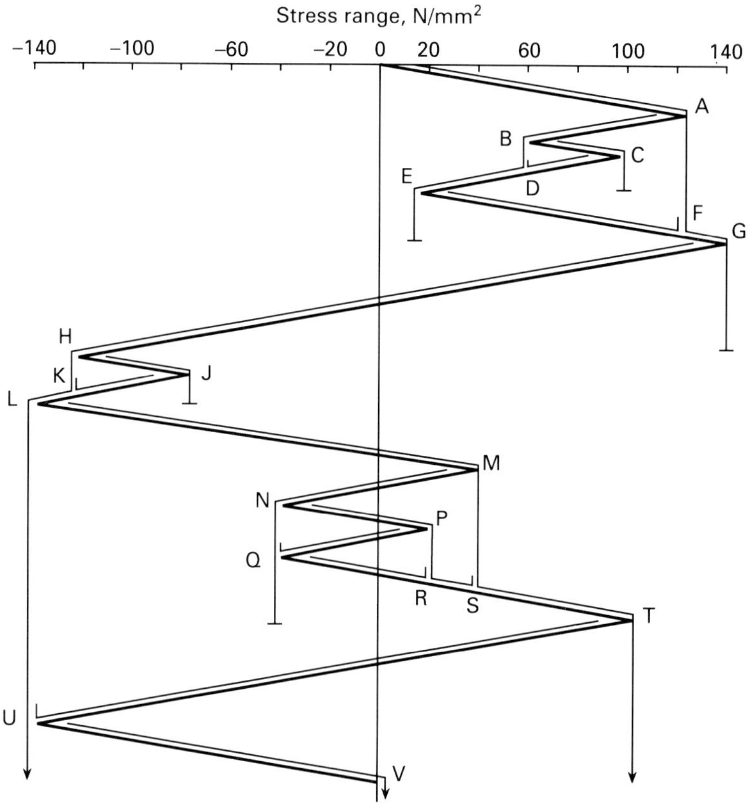

Results:

Stress range, N/mm^2	Positive half-cycles	Negative half-cycles	Total cycles
20			
40	11	11	2
60	1	1	1
80	1	1	1
100	1	1	1
120			
140	11		$\frac{1}{2}$
160			
180			
200			
220			
240	1	1	1
260			
280		1	$\frac{1}{2}$

AII.2 Rainflow counting method.

To use the method it is necessary to present the waveform in such a way that it starts and finishes at the same peak value, which is also the maximum peak. The waveform used previously, labelled in the same way as for the rainflow example, in which the maximum peak occurs only once has been redrawn, Fig.AII.3, to fulfil this condition by starting at G, the maximum peak, and adding sequence 0–G after point V has been reached. Such a modification to the waveform will always be necessary unless the stress history actually starts and finishes at the maximum peak levels.

The waveform is imagined to be the profile of a reservoir, initially full of water, which has plugs at each trough level. Cycles are counted by observing how the reservoir is emptied when water is drained from the troughs:

- Remove the plug from the lowest trough and allow water to drain. If there are two or more such troughs, as in the example shown (L and U), drain from them one at a time in any order. In the example, the reservoir is first drained from L. When the first draining operation is complete, list one cycle having a stress range equal to the vertical height of water drained, that is 280 N/mm² in the example.
- Repeat the operation by draining from the other plugs at the same level, that is U in the example, and then successively at higher trough levels (H, N, Q, E, and B in the example). The shaded areas in the Figure show those parts of the reservoir which have been drained at the end of each operation. List one cycle at the end of each draining, as indicated in the Table in Fig.AII.3.

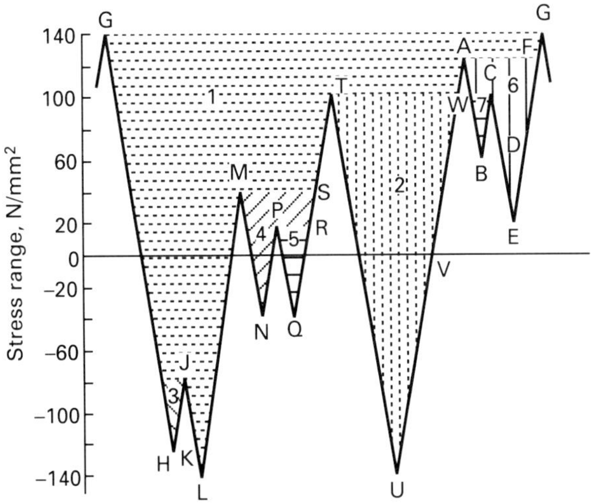

Stress range, N/mm²	No. of cycles	Draining operation
280	1	1
260		
240	1	2
220		
200		
180		
160		
140		
120		
100	1	6
80	1	4
60	1	5
40	2	3, 7
20		

AII.3 Reservoir counting method.

It may be noted that the reservoir and rainflow methods give the same result if the waveform is considered in the form used to carry out the reservoir count. In the example, the two positive half-cycles of range 140 N/mm^2 would combine to give one positive half-cycle of range 280 N/mm^2 if the waveform is considered as going from G to E, rather than from 0 to V.

Although cycle counting techniques may seem complicated, they can readily be automated by computer. Indeed, with the aid of modern data logging equipment and associated computers, cycle counting can become just one of the steps in a fully-automated analysis of service loading measurements made on actual structures.

Appendix III

Fracture mechanics analysis of fatigue

Introduction

Fracture mechanics is concerned with cracks and how they behave under load. Fracture mechanics principles are applicable only to fatigue if a crack or a crack-like flaw is present. Thus, for example, they are applicable to the consideration of:

- Fatigue cracks.
- Small inherent crack-like flaws at the toes of welds.
- Planar flaws which may or may not be real cracks, (e.g. hydrogen cracking, solidification cracking, lack of fusion, incomplete penetration).
- Non-planar flaws (e.g. inclusions) can also be considered, treating them as cracks. However, fracture mechanics only considers the fatigue crack propagation part of the fatigue life, which will be less than the actual life if there is a significant fatigue crack initiation period. This might be the case with some non-planar flaws.

In the context of fatigue, fracture mechanics is used primarily to describe the rate at which a crack will propagate under repeated loading, but it can also be used to determine the extent of fatigue cracking which can be tolerated before fracture occurs, i.e. to define the critical crack corresponding to the material's fracture toughness. Only its application to fatigue crack propagation is dealt with here.

Stress intensity factor

The strength of fracture mechanics lies in the fact that general relationships can be developed which are applicable to cracks and components with a variety of geometries. This is achieved by using a parameter K, called the stress intensity factor, which is proportional to the actual stress in the elevated stress field developed around the tip of a crack in a member under load. Calculated by linear elastic stress analysis, K can refer to one of three possible modes of loading on a crack,

Fig.AIII.1. The crack opening mode, mode I, is the most widely encountered in practice. In general

$$K = Y\,S\sqrt{\pi a} \qquad [AIII.1]$$

where a is the crack length, S is the nominal stress acting normal to the crack and Y is a function which depends upon the geometry of the crack and cracked part under consideration. Many 'solutions' for K exist and are documented in the literature, the relevant information usually being the term Y as a function of crack size. It is important to define 'a' correctly: it is the depth of a surface or edge but only half the depth or length of a buried or through-thickness crack, Fig.AIII.2.

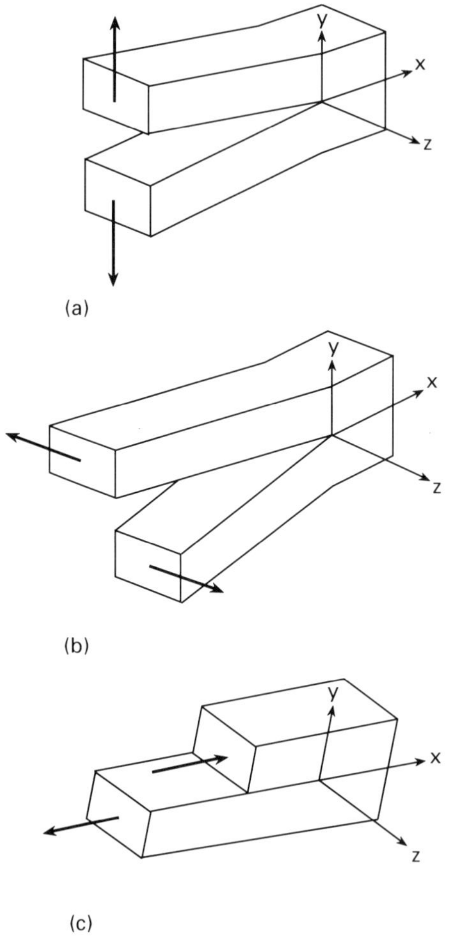

AIII.1 Modes of crack surface displacement: (a) I, stress intensity factor = K_I; (b) II (K_{II}); (c) III (K_{III}).

Fatigue crack propagation

The basis of the application of fracture mechanics to fatigue is the relationship between rate of crack propagation $\frac{da}{dN}$, Fig. AIII.3a, and stress intensity factor range, ΔK, based on applied cyclic stress range, which for many practical cases has the simple form:

$$\frac{da}{dN} = C\,(\Delta K)^m \qquad \text{[AIII.2]}$$

where C and m are material constants which are found experimentally by measuring the rate of growth of a crack under known loading in a simple test specimen, Fig.AIII.3b. At high values of the maximum

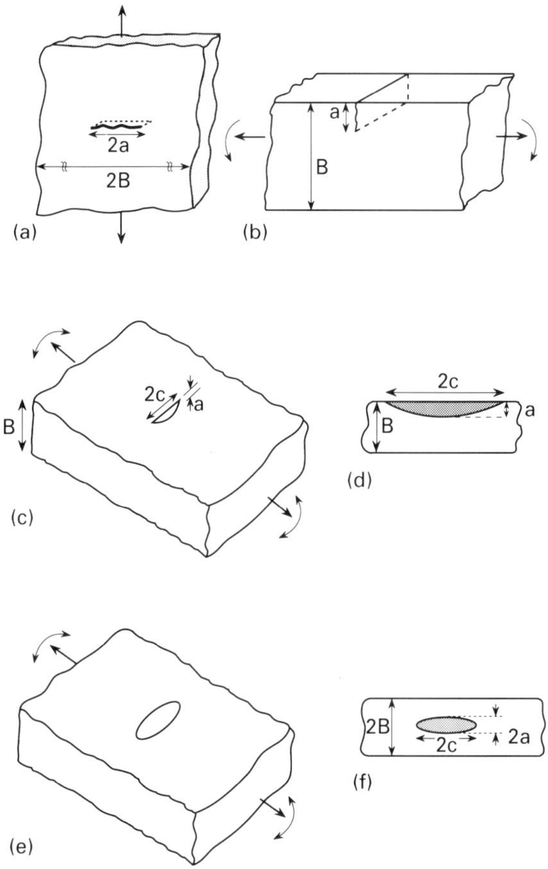

AIII.2 Crack dimensions: (a) Through-thickness crack; (b) Edge crack; (c) Surface crack; (d) Section through plane of crack; (e) Buried crack; (f) Section through plane of crack.

stress intensity factor (corresponding to the peak stress in the cycle) towards the end of the life of the crack, there is an acceleration in rate of crack propagation as the critical value of K for unstable fracture is approached. However, in most circumstances crack size increases exponentially with endurance, with the result that this behaviour occupies an insignificant part of the fatigue life and can therefore normally be ignored. More important (see Appendix I), at low values a threshold value of ΔK, ΔK_0, is approached below which cracks will not propagate. For a given geometry of cracked component (e.g. a particular type of welded joint with inherent weld toe flaws) ΔK_0 determines the endurance limit.

The experimentally determined values of m and C vary but for struc-

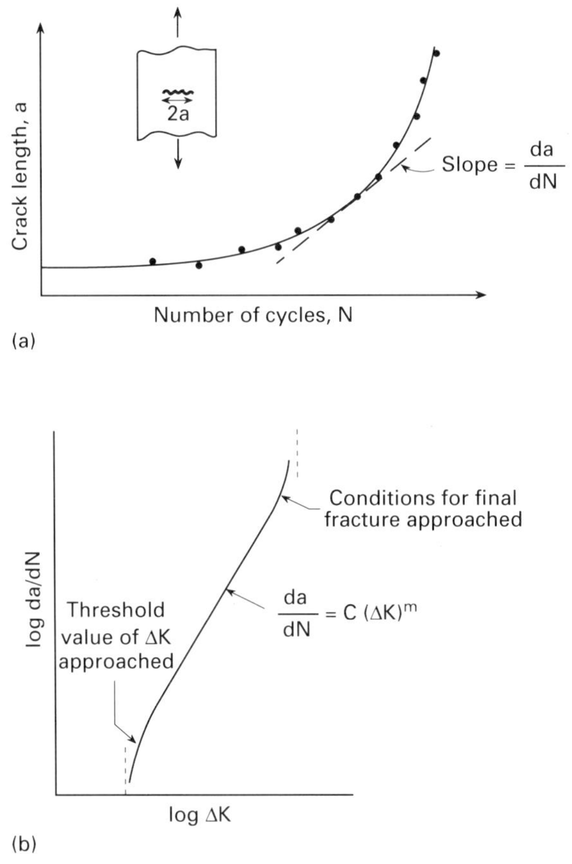

AIII.3 Fatigue crack propagation: (a) Basic data; (b) Fracture mechanics representation.

tural steel and weldable aluminium alloys the most widely used value for m is 3, Fig.AIII.4. Fatigue crack propagation data for different materials correlate reasonably well when plotted in terms of $\Delta K/E$, where E is the elastic modulus. As seen in Fig.AIII.4, this is the situation for steel and aluminium alloy whose modulii differ by a factor of 3. Corresponding values of C representing conservative estimates of rate of crack propagation (excluding environmental effects) are 3×10^{-13} and 8×10^{-12} for steel and aluminium respectively (for units of N/mm^2 and millimetres).

Integration of fatigue crack propagation law

Substituting for ΔK in [AIII.2] gives

$$\frac{da}{dN} = C\,(YS\,\sqrt{\pi a})^m \qquad \text{[AIII.3]}$$

This relationship becomes more useful after re-arranging it to bring together the terms which depend upon 'a' and integrating between limits:

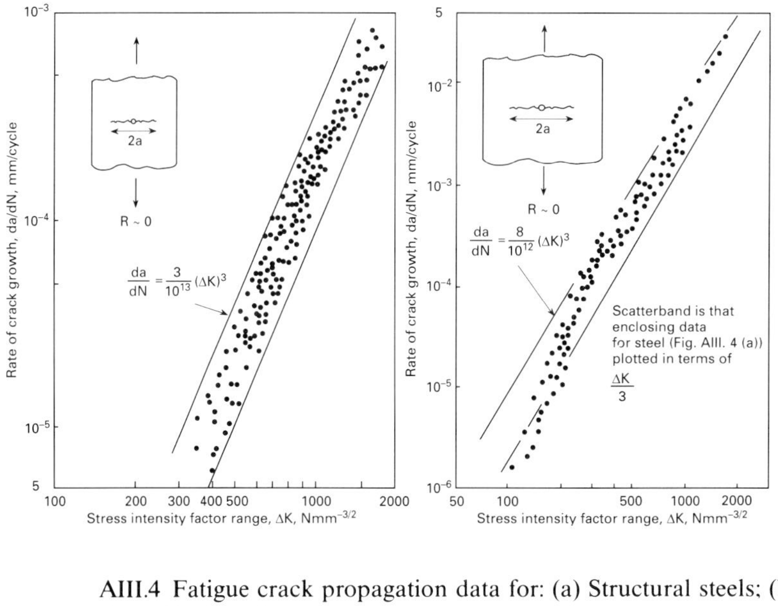

AIII.4 Fatigue crack propagation data for: (a) Structural steels; (b) (b) Al-Zn-Mg-alloy.

$$\int_{a_{init}}^{a_f} \frac{da}{(Y\sqrt{\pi a})^m} = CS^m \int_0^N dN = CS^m N \qquad [AIII.4]$$

Thus, an initial crack of size a_{init} is calculated to propagate to a size of a_f in N cycles under stress range S. This equation represents a major step forward in fatigue analysis because it allows the fatigue process to be considered at any stage in the fatigue life, in contrast to the S–N curve which relates to one event: final failure. The flexibility it introduces is such that any initial flaw or crack size (value of a_{init}) can be considered and any failure criterion (value of a_f) can be incorporated. Thus, a_{init} might refer to a detected fatigue crack or a planar weld flaw while a_f might be critical crack size, (e.g. readily detectable, fracture,) or the attainment of through-thickness cracking, (e.g. for leakage of a vessel). Since rate of crack growth usually increases as the crack size increases, much of the fatigue life may be exhausted with little increase in crack size, Fig.AIII.3. Therefore, as long as a_f is very much greater than a_i its actual value has little influence on the fatigue life of the original cracked part.

In practice, it is often convenient to express [AIII.4] in terms of a/B, where B is the section size (see Fig.AIII.2 for definitions of B), especially as Y is a function of a/B. Then

$$\int_{a_{init}/B}^{a_f/B} \frac{d\,(a/B)}{\left(Y\sqrt{\frac{\pi a}{B}}\right)^m} = C\,S^m\,B^{\left(\frac{m}{2}-1\right)}N \qquad [AIII.5]$$

Calculating the S–N curve

It is of interest to use [AIII.4] to consider the fatigue behaviour of a nominally flaw-free as-welded joint in which failure consists of the growth of a crack from an inherent cracklike flaw, (e.g. weld toe failure). In this case, if the geometry, including size of inherent flaw, remains the same from one weld to the next, as is assumed for welded joints tested to generate S–N data, and the crack size at failure is very large compared with a_{init}, the integral in [AIII.4] is constant and the equation becomes

$$S^m N = \text{a constant} \qquad [AIII.6]$$

This is, of course, the form of S–N curve actually obtained from fatigue tests of welded joints. In reality there are variations in geometry from one weld to the next so that the integral in [AIII.4] is not really constant. Such variations contribute to scatter observed in fatigue test data.

It may be noted that, as well as predicting the correct form of the S–N relationship, the fracture mechanics analysis also gives the slope of the S–N curve m, which, as will be seen, should be the same as the slope of the crack propagation data for the material, Fig.AIII.5. For structural steels, as noted above, m is usually around 3, Fig.AIII.4, which is the reason for the choice of m=3 for the slopes of the design S–N curves for Classes D to W, covering details in which fatigue cracks propagate from pre-existing cracklike flaws. Such a slope was compatible with basic analysis of the test data assuming an equation of the form of [AIII.6]. Joints of higher classification (B, C) do not fail from cracklike flaws and a significant part of their fatigue lives may be occupied initiating a fatigue crack. Then, shallower S–N curves are obtained.

Crack propagation under variable range loading

To consider the growth of a crack under variable stress range loading, [AIII.4] (or [AIII.5]) is applied to successive stress cycles. Thus, if the load spectrum produces n_1 cycles at stress range S_1, n_2 cycles at S_2, etc

$$\int_{a_{init}}^{a_1} \frac{da}{(Y\sqrt{\pi a})^m} = C\, S_1^{\,m} n_1$$

$$\int_{a_1}^{a_2} \frac{da}{(Y\sqrt{\pi a})^m} = C\, S_2^{\,m} n_2 \qquad \text{etc}$$

or, over the total life of the crack, that is until the crack reaches the predetermined critical size, a_f

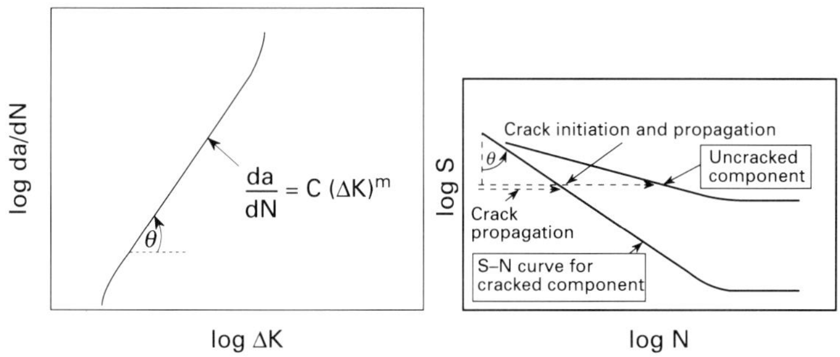

AIII.5 Relationship between crack propagation law and S–N curve.

$$\int_{a_{init}}^{a_1} \frac{da}{(Y\sqrt{\pi a})^m} + \int_{a_1}^{a_2} \frac{da}{(Y\sqrt{\pi a})^m} + \ldots\ldots\ldots\ldots + \int_{a_j}^{a_f} \frac{da}{(Y\sqrt{\pi a})^m} = C\,(S_1{}^m n_1 + S_2{}^m n_2 + \ldots\ldots S_k{}^m n_k) \quad \text{[AIII.7]}$$

at failure, which is actually equivalent to Miner's rule. This can be proved by deriving Miner's Rule from the above, as follows.

The sum of the integrals on the left-hand side of [AIII.7] is the same as the integral from $a = a_{init}$ to a_f which appears in [AIII.4]. Calling the integral I, [AIII.4] can be used to express the stresses in [AIII.7] in terms of the corresponding constant amplitude fatigue lives at those stresses, i.e.

$$S_1{}^m = \frac{I}{CN_1}, \quad S_2{}^m = \frac{I}{CN_2}, \quad \text{etc.}$$

Equation AIII.7 can thus be expressed

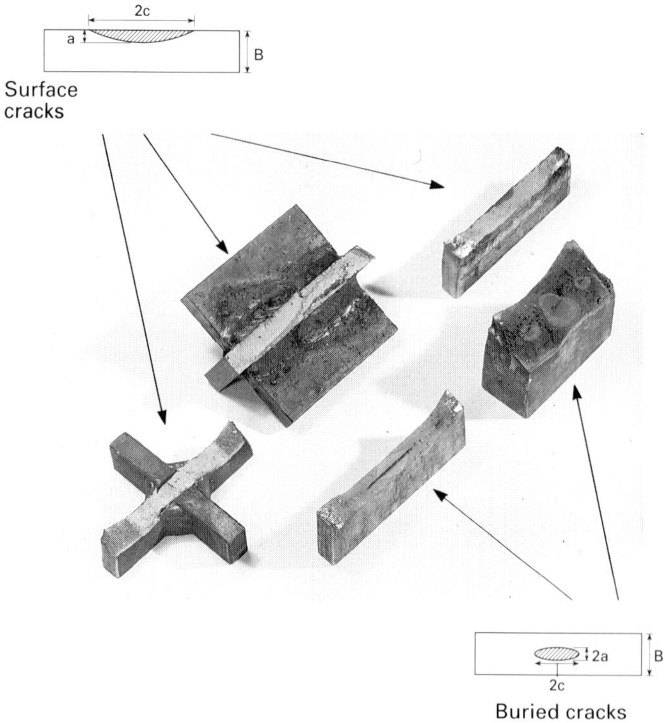

AIII.6 Shape of fatigue cracks in some welded joints.

$$I = C\left[\frac{I}{CN_1} \cdot n_1 + \frac{I}{CN_2} \cdot n_2 + \frac{I}{CN_3} \cdot n_3 + \text{etc}\right] = I\sum \frac{n_i}{N_i}$$

[AIII.8]

Where i = 1, 2, 3, etc, i.e.

$$\sum \frac{n_i}{N_i} = 1$$

at failure. as stated by Miner's rule.

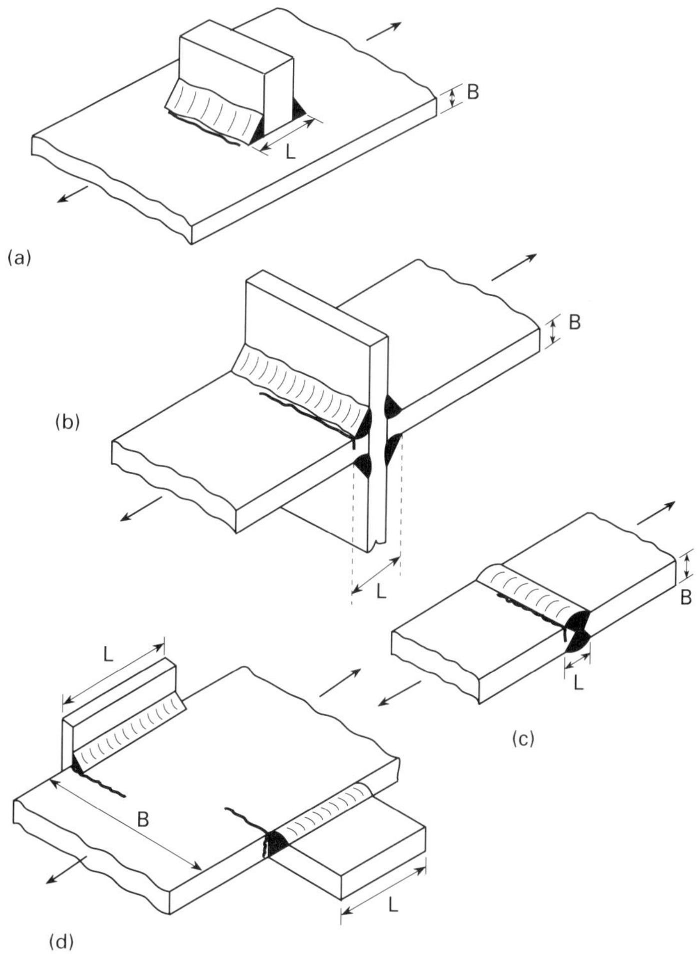

AIII.7 Definitions of 'L' and 'B' for various welded joints.

Cracks in welded joints

Fatigue cracks which occur in welded joints are of the surface or embedded types, Fig.AIII.6, so that the crack front has an elliptical shape. In these circumstances, Y, must incorporate the correction for crack front shape a/2c as well as a/B. In general, Y increases with increase in a/B and decrease in a/2c. In other words, the most severe shape of crack is one with a straight crack front, a/2c = 0.

For cracks which propagate in the region of stress concentration produced by the weld profile, that is cracks at the weld toe, a further correction, M_K, is introduced (NB: as noted earlier, it is not necessary to take account of this stress concentration in the determination of design stresses). Stress analysis of weld toe cracks in a range of profiles representing fillet and butt welded joints has revealed a relationship between M_K, crack size and welded joint geometry. In particular, it is found that M_K, strongly depends on the joint proportions ratio L/B (Fig.AIII.7), where L is the 'attachment' length. Here the word 'attachment' is being used to describe the length (i.e. dimension in the direction of stressing) over which the main section thickness (B) is increased as a result of the weld(s). Thus, it might refer to a genuine attachment, Fig.AIII.7a, to a cruciform joint, Fig.AIII.7b, or to a transverse butt weld bead, Fig.AIII.7c. A fourth example is the case of an attachment on the edge of a stressed plate, Fig.AIII.7d. In this case, the dimension B is actually equivalent to the width rather than its thickness. The weld toe profile, both angle of incidence and toe radius, is also significant, but only for very small cracks. The general form of the relationship between M_K, a/B and L/B is:

$$M_K = \alpha \left(\frac{a}{B}\right)^{\beta} \qquad \text{[AIII.9]}$$

where the constants α and β are both functions of L/B. The values of α and β, together with other stress intensity factor solutions relevant to fatigue cracks in welded joints, are included in BS PD 6493 (see Chapter 4).

A particular feature of M_k for weld toe cracks is that it decreases with increase in a/B, as the crack tip moves further away from the region of stress concentration, until it reaches unity, typically 25–30% of the way through the section. An important consequence of this is that for a given crack depth the stress intensity factor of a crack in a thick plate is higher than that in a thin plate, as shown schematically in Fig.AIII.8 for a surface crack at the toe of a weld. The result is that the fatigue strength of the thin member is higher than that of the thick. This will also be true for welded joints as fatigue cracks in these initiate at small pre-existing cracklike flaws whose depths do not depend upon plate thickness and therefore could be similar in thin and thick plates. Test results which show the influence of plate thickness for joints failing from the weld toe were presented earlier. It may be noted that, in the absence of a stress concentration, i.e. $M_k = 1$, for a given crack depth, Y is less in the thick plate than in the thin one, with the result that fatigue strength will increase with thickness.

Practical applications

Practical applications of fracture mechanics fall into two main categories:
(a) Calculation of fatigue performance.
(b) Parametric studies.

Calculation of fatigue performance is any application of equations [AIII.4] or [AIII.5] with one of the variables as an unknown. Thus, examples include:

- Assessment of welding flaws (see Chapter 4).
- Estimation of tolerable flaw size for specified fatigue strength (a_{init} unknown).
- Calculation of remaining life of a fatigue crack detected in service (N unknown).
- Calculation of inspection periods during which a critical crack will not develop (N unknown).
- Making allowance for material fracture toughness in fatigue performance calculations (N or S and N (i.e. the S–N curve) unknown).

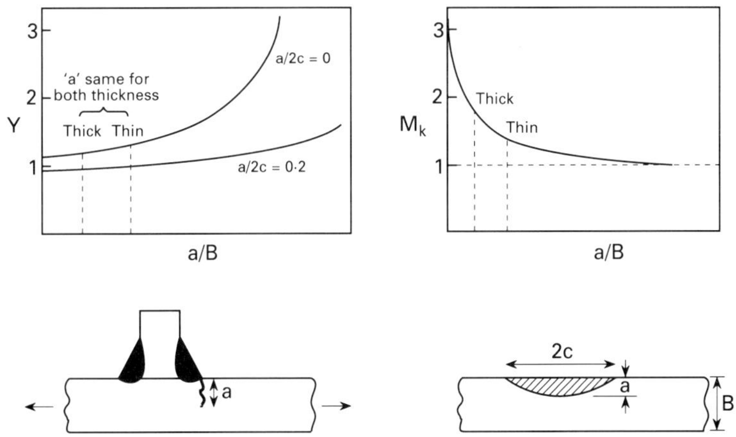

AIII.8 Influence of plate thickness stress intensity factor for crack of given depth at toe of weld.

The accuracy of fatigue performance calculations is greatly dependent on the accuracy of the input data. For example, scatter in fatigue crack growth data for a given material can amount to a factor of 4 on da/dN and hence on C in [AIII.2], which in turn leads to errors in fatigue life estimates by a factor of 4. The wrong choice of m in [AIII.2] can also be significant since N is proportional to S^m. However, of even greater significance is errors in S, since m is usually of the order of 3. Thus, an underestimation of S by 40% leads to an overestimate in life of 4 times. Finally, there will be errors in the estimates of the dimensions of the initial flaw and critical crack. The former is generally more significant, as da/dN increases exponentially. For example, for m = 4 the crack spends approximately half its life doubling in size. The proportion is lower for lower m values or if the crack is propagating in a region of decreasing local stress concentration, but still significant. Flaw shape is also significant as ΔK increases with decrease in flaw aspect ratio, the most severe shape being a straight fronted crack. Limitations in non-destructive inspection capabilities mean that flaw dimensions are rarely known accurately. Clearly, every effort should be made to establish the input data as accurately as possible but, in the absence of accurate data, it would be necessary to make conservative assumptions about their values.

The use of [AIII.4] or [AIII.5] for **parametric studies** is less susceptible to errors in the input data, as the equations are used only for comparisons. In general, the equations are used to explore the influence of *changes* to the input parameters. Examples include:

- To compare the fatigue strengths of joints in different materials or environments (vary C and m).
- To determine the significance from the fatigue viewpoint of material toughness (vary a_f).
- To explore the effect of changes to the weld joint geometry (vary B and/or L and hence M_k).

Such calculations can even be used to extend a limited fatigue database, by using known fatigue performance results (e.g. obtained from one particular geometry of welded joint) to 'calibrate' [AIII.4] or [AIII.5]. In particular, values of m and a_{init} can be deduced from the S–N curve and then used to assess the change in the S–N curve which would result from changes in, say, plate thickness.

Bibliography

Fatigue

General

Collins JA: 'Failure of materials in mechanical design'. New York, John Wiley & Sons, 1981.

Forrest PG: 'Fatigue of metals'. Oxford, Pergamon Press, 1962.

Frost NE, Marsh KJ, and Pook LP: 'Metal fatigue'. Oxford, Clarendon Press, 1974.

Fuchs H and Stephens RI: 'Metal fatigue in engineering'. New York, John Wiley & Sons, 1981.

Welded joints

Gurney TR: 'Fatigue of welded structures', 2nd/ed, Cambridge, University Press, 1979.

Gurney TR: 'Fatigue tests on fillet welded joints to assess the validity of Miner's cumulative damage rule'. *Procs R Soc London, A386*, 1983, 393–408.

Fatigue data for welded joints

Gurney TR and Maddox SJ: 'A re-analysis of fatigue data for welded joints in steel'. *Welding Research Int* 1973 3 (4) 1–54.

Olivier R and Ritter W: 'Catalogue of S–N curves of welded joints in structural steels, Pts 1–5'. Dusseldorf, Deutscher Verband für Schweisstechnik, 1979–82.

Sanders WW Jr: 'Fatigue behaviour of aluminium alloy weldments'. *WRC Bull* 1972, 171, and 1983, 286. New York, Welding Research council.

NRIM fatigue data sheets, National Research Institute for Metals, Tokyo, Japan.

Fatigue design rules in British codes and specifications

BS 5400: Pt 10: 1980 'Specification for steel, concrete and composite bridges. Code of practice for fatigue', British Standards Institution, London.

BS 5500 1988 'Specification for unfired fusion welded pressure vessels, Enquiry Case No. 79', British Standards Institution, London.

BS 8118: 1991 'The structural use of aluminium', British Standards Institution, London.

BS PD 6493: 1991 'Guidance on methods for assessing the ability of flaws in welded structures', British Standards Institution, London.

Offshore Installations: Guidance on design and construction, Department of Energy, HMSO, 1990.

International fatigue design codes and specifications

'IIW Design recommendations for cyclic loaded welded structures', *Welding in the World* 1982 20 (728) 153–165.

Eurocode No.3: Common unified rules for steel structures, EUR 8849, Commission of the European Communities, Luxembourg, 1984.

'IIW Guidance on assessment of the fitness for purpose of welded structures', International Institute of Welding, Cambridge, SST-1157-90, ISBN 876-88771-77-6, 1990.

Improvement methods

Booth GS (Ed): 'Improving the fatigue performance of welded joints'. Abington, Welding Institute, 1983.

Fatigue failures

Anon: 'Fatigue fractures in welded constructions'. Vol 1 (1967), Vol 2 (1979). Paris, Int'l Institute of Welding.

Fracture mechanics

Broek D: 'Elementary engineering fracture mechanics' Noordhof International Publishing, 1974.

Anon: 'A general introduction to fracture mechanics'. Mechanical Engineering Publications, 1975.

Rolfe ST and Barsom JM: 'Fracture and fatigue control in structures, application of fracture mechanics'. New Jersey, Prentice-Hall, 1977.

Rooke DP and Cartwright DJ: 'Compendium of stress intensity factors'. London, HMSO, 1974.

Index

Illustrations are referenced in italic